Multifunctional Polymeric Foams

Polymeric foams or cellular or expanded polymers have characteristics that make their usage possible for several industrial and household purposes. This book is focused on the recent advancements in the synthesis of polymer foams, various foaming methods, foaming technology, mechanical and physical properties, and the wide variety of its applications. Divided into 11 chapters, it explains empirical models connecting the geometrical structure of foams with their properties including structure-property relations.

This book:

- Describes functional foams, their manufacturing methods, properties, and applications.
- Covers various blowing agents, greener methods for foaming, and their emerging applicability.
- Illustrates comparative information regarding polymeric foams and their recent developments with polymer nanocomposite foams.
- Includes applications in mechanical, civil, biomedical, food packaging, electronics, health care industry, and acoustics fields.
- Reviews elastomeric foams and their nanocomposite derivatives.

This book is aimed at researchers and graduate students in materials science, mechanical engineering, and polymer science.

Emerging Materials and Technologies

Series Editor: Boris I. Kharissov

The *Emerging Materials and Technologies* series is devoted to highlighting publications centered on emerging advanced materials and novel technologies. Attention is paid to those newly discovered or applied materials with potential to solve pressing societal problems and improve quality of life, corresponding to environmental protection, medicine, communications, energy, transportation, advanced manufacturing, and related areas.

The series takes into account that, under present strong demands for energy, material, and cost savings, as well as heavy contamination problems and worldwide pandemic conditions, the area of emerging materials and related scalable technologies is a highly interdisciplinary field, with the need for researchers, professionals, and academics across the spectrum of engineering and technological disciplines. The main objective of this book series is to attract more attention to these materials and technologies and invite conversation among the international R&D community.

Nanotechnology Platforms for Antiviral Challenges
Fundamentals, Applications and Advances
Edited by Soney C George and Ann Rose Abraham

Carbon-Based Conductive Polymer Composites
Processing, Properties, and Applications in Flexible Strain Sensors
Dong Xiang

Nanocarbons
Preparation, Assessments, and Applications
Ashwini P. Alegaonkar and Prashant S. Alegaonkar

Emerging Applications of Carbon Nanotubes and Graphene
Edited by Bhanu Pratap Singh and Kiran M. Subhedar

For more information about this series, please visit: www.routledge.com/Emerging-Materials-and-Technologies/book-series/CRCEMT

Multifunctional Polymeric Foams

Advancements and Innovative Approaches

Edited by
Soney C. George and Resmi B. P.

CRC Press
Taylor & Francis Group
Boca Raton London New York

CRC Press is an imprint of the
Taylor & Francis Group, an **informa** business

First edition published 2023
by CRC Press
6000 Broken Sound Parkway NW, Suite 300, Boca Raton, FL 33487-2742

and by CRC Press
4 Park Square, Milton Park, Abingdon, Oxon, OX14 4RN

CRC Press is an imprint of Taylor & Francis Group, LLC

ISBN: 9781032111698 (hbk)
ISBN: 9781032111704 (pbk)
ISBN: 9781003218692 (ebk)

DOI: 10.1201/9781003218692

Typeset in Times
by codeMantra

Contents

Chapter 3   Latex-Based Polymeric Foams – Preparation, Properties and Applications 41

Leny Mathew

Chapter 4   Blended Polymeric Foams 51

Tharun A. Rauf and Athira C. J.

Chapter 7   Polymer Nanocomposite Foams and Acoustics 111

Benjamin Tawiah, Charles Frimpong and Bismark Sarkodie

Editor Biographies

Dr. Soney C. George is the Dean Research and Director at the Centre for Nanoscience and Technology, Amal Jyothi College of Engineering, Kanjirappally, Kerala, India. He is a Fellow of the Royal Society of Chemistry, London, and a recipient of "best researcher of the year 2018" award from APJ Abdul Kalam Technological University, Thiruvananthapuram, India. He has also received awards such as best faculty award from the Indian Society for Technical Education, best citation award from the International Journal of Hydrogen Energy, a fast-track award of young scientists by the Department of Science & Technology, India, and an Indian Young Scientist Award instituted by the Indian Science Congress Association. He is also a recipient of CMI Level 5 Certificate in Management and Leadership from Dudley College of Technology, London as part of AICTE-UKIERI Training Programme. He did his postdoctoral studies at the University of Blaise Pascal, France, and Inha University, South Korea. He has guided eight PhD scholars and 102 student projects. He has published and presented almost 240 publications in journals and conferences and his h-index is 27. His major research fields are polymer nanocomposites, polymer membranes, polymer tribology, pervaporation, and supercapacitors. Besides he took leadership in organizing a sponsored international conference on Engineering Education in association with IEEE International chapter, as well as conferences on development of Nanoscience and Technology. He initiated several lecture series like Isaac Newton Lecture Series, ACeNT Lecture Series, and Lecture Series on Nobel Prize Winning Works in order to bring eminent scientists and academicians from India and abroad to the College. Scientists from USA, Malaysia, France, Poland, and South Africa have visited the campus and interacted with students through his networking. Collaborative research work was also initiated between the Gdansk University of Technology, Poland; Inha University, South Korea; University of Blasé Pascal, France; Centre for Nanostructures and Advanced Materials, CSIR, South Africa; Durham University, United Kingdom.

Resmi B. P. received her MSc degree from Kerala University, India. She is also having another postgraduate degree in education from Kerala

University. She is currently working as Assistant Professor at the TKM Institute of Technology, Kollam, Kerala, India in the Department of Science and Humanities. She has ten years of experience in teaching chemistry as well as in education. Now she is doing PhD in the area of Polymeric Foams. She has authored a chapter in the book *Applications of Multi Functional Nanomaterials* by Elsevier.

List of Contributors

Vipul Agarwal
Cluster for Advanced Macromolecular Design (CAMD), School of Chemical Engineering, University of New South Wales, Sydney, NSW 2052, Australia

Aswathy R.
TKM Institute of Technology, Ezhukone, Kollam 691505, Kerala, India

Athira C. J.
Mannam Memorial NSS College, Kottayam, Kollam 691571, Kerala, India

Souhardya Bera
Department of Chemical Engineering, University of Calcutta, 92 A. P. C. Road, Kolkata 00009, India

Pradeep Bhaskar
Department of Physics, School of Engineering, Presidency University, Bengaluru 560064, Karnataka, India

Rishov Kumar Das
Department of Chemical Engineering, University of Calcutta, 92 A. P. C. Road, Kolkata 00009, India

Arvil Dasgupta
Department of Chemical Engineering, University of Calcutta, 92 A. P. C. Road, Kolkata 00009, India

Charles Frimpong
Department of Industrial Art (Textiles), Kwame Nkrumah University of Science and Technology, PMB, Kumasi, Ghana

Jinu Jacob George
Department of Polymer and Rubber Technology, Cochin University of Science and Technology, Kochi 682022, Kerala, India

Soney C. George
Centre for Nanoscience and Technology, Amal Jyothi Engineering College, Koovapally, Kerala, 686518, India

Jobin Jose
Interdisciplinary Research Centre in Construction and Building Materials, Research Institute, King Fahd University of Petroleum Minerals, Dhahran 31261, Saudi Arabia

Behrad Koohbor
Department of Mechanical Engineering, Rowan University, Glassboro, NJ, USA

Leny Mathew
Mahatma Gandhi University College of Engineering, Muttom, Thodupuzha 685587, Kerala, India

Nizam P. A.
School of Chemical Science, Mahatma Gandhi University, Kottayam, Kerala, India

Tharun A. Rauf
TKM College of Arts and Science,
  Karicode, Kollam 691005, Kerala,
  India

Ratheesh P. M.
Department of Instrumentation,
  Cochin University of Science and
  Technology, Kochi-682022, Kerala,
  India.

Resmi B. P.
TKM Institute of Technology,
  Ezhukone, Kollam, 691505, Kerala,
  India

Arnab Roy
Department of Chemical Engineering,
  University of Calcutta, 92 A. P. C.
  Road, Kolkata 00009, India

Subhasis Roy
Department of Chemical Engineering,
  University of Calcutta, 92 A. P. C.
  Road, Kolkata 00009, India

Anu Sasi
Department of Polymer and Rubber
  Technology, Cochin University of
  Science and Technology, Kochi
  682022, Kerala, India

Bismark Sarkodie
Key Laboratory of Ultrafine Materials
  of Ministry of Education, School of
  Materials Science and Engineering,
  Shanghai, China

Ranimol Stephen
Department of Chemistry, St. Joseph's
  College (Autonomous), Devagiri,
  Calicut 673 008, Kerala, India.
  Email: ranistephen@gmail.com

Benjamin Tawiah
Department of Industrial Art (Textiles),
  Kwame Nkrumah University of
  Science and Technology, PMB,
  Kumasi, Ghana. Email: tawiahb@
  gmail.com

Sabu Thomas
International and Inter-University
  Centre for Nanoscience and
  Nanotechnology, Mahatma Gandhi
  University, Kottayam, Kerala, India

George Youssef
Experimental Mechanics Laboratory,
  Mechanical Engineering
  Department, San Diego State
  University, San Diego, CA, USA

Preface

Polymeric foams continue to grow at a rapid pace throughout the world. Polymer foam is an important polymer material that exhibits low density, good heat and insulation effects, high-specific strength, and resistance to corrosion. Polymer foams can be rigid, flexible, or elastomeric, and can be produced from a wide range of polymers. Thus, enhanced properties like lightweight, high impact strength, and improved insulation properties make polymer foam find enormous applications. The cell size and the morphology of the foam in turn lead to enhancement of its properties resulting in microcellular and nanocellular foams. The combination of lightweight and high strength determines the quality of polymer foams. One can easily find a lot of published information on polymeric foams, their processing strategies, blowing agents, and their various applications depending upon the property enhancement. Therefore, the central theme of the proposed book is to review in a fairly comprehensive manner the recent accomplishments in polymeric foaming technologies, blowing agents, and applications.

In this edited book an overall view of the status of polymeric foams is presented in a lucid style. The first chapter narrates about the different types of polymeric foams and their potential applications. The second chapter discusses the micromechanical modeling of foams, and the third chapter narrates the preparation, properties, and applications of latex-based polymeric foams. Chapter 4 deals with polymer blend foams while Chapter 5 provides an insight into the role of polymer nanocomposite foams as metal ion removers. The application of polymeric nanocomposite foams in biomedical areas is summarized in Chapter 6. Polymer nanocomposite foams in acoustics applications are discussed in Chapter 7. Current trends in the use of elastomeric foams and their nanocomposite derivatives in the foam market are examined in Chapter 8.

Chapter 9 provides the role of polymeric foams and their nanocomposite derivatives in shock absorption areas. Chapter 10 gives an overview of emerging applications of polymeric nanocomposite foams, whereas Chapter 11 summarizes the innovations of polymeric foams and their new application opportunities including energy and energy devices.

The editors are thankful to all distinguished contributors of this book. We are also grateful to the reviewers of chapters for their wonderful help. Our sincere gratitude to CRC press and the publishing team for their wholehearted support to complete this project. Our sincere gratitude to the Manager, Director, Principal, and Members of the staff of Amal Jyothi College of Engineering and ACeNT for their constant encouragement and support.

1 Polymeric Foams – An Introduction

Resmi B. P. and Soney C. George

CONTENTS

1.1 INTRODUCTION

Innovations and developments in technology enhance the production of polymeric foams as useful products in day-to-day life. It is found that low-density polymeric foams have been produced before World War II. There was a steady growth in the production of polymeric foams based on a study reported in 2019 as more than 50 million lb. It is due to the developments in science and technology that leads to the

DOI: 10.1201/9781003218692-1

changes in processing methods. Due to this, novel and innovative resources have been utilized in foaming technology. Nanotechnology and its faster developments also play important roles in the production of porous polymeric materials with different properties and a variety of applications, including sports, furniture, food segments, construction, aerospace, and automotive. The advances in the use of nanomaterials as fillers in polymeric materials also lead to the development of nanocomposite foams, which makes porous polymeric materials as one of the important useful products in the industry (Khemani, 1997). Nowadays, polyurethane (PU) and polystyrene foams (PS), emerge as top foam products with a wide variety of applications. PS foam was synthesized in the year 1930 as it was the first foamed product. Familiar examples are sofa cushions or chair cushions, or car seats made of PU foam. Coffee or tea cup used in restaurants are made of PS foam. Nowadays, many researchers are focusing on the synthesis of polymeric foams to produce lightweight nanocomposite foams having applications in tissue engineering, scaffolds, electrical as well as thermal insulation, membranes in the separation process, etc. Moreover, multifunctional polymeric foams having carbon nanoparticles emerged in more interest in the industry due to density reduction by the process of foaming. Based on the pore size, polymeric foams can be classified into macrocellular foams (typical cell sizes of 50 μm or larger), MCFs (typical cell sizes of 10–100 μm), and Nanocellular foams (typical cell sizes of below 100 nm. Polymeric foams, in general, are composed of small cavities inside a polymer matrix and this produces good sound and thermal insulation, resistance to corrosion, high-specific strength, low density, etc. Foam injection molding and foam extrusion molding are the two main fabrication methods in polymer foaming. Apart from PU and PS foams, there are polypropylene foams extensively used in, military industries, aerospace, daily necessities, and transportation for instance in the manufacture of many seat cushions, packaging materials, shockproof materials, building materials, and thermal insulation materials. Phenolic foam generally has applications in the fields of automobiles, architecture, electrical and electronic applications, the iron and steel industry, and aerospace. It can meet the requirements of heat insulation and combustion-resistant building materials for large-scale sports and entertainment places, high-rise buildings, and high-speed vehicles. Polyimide (PI) foam, having high thermal insulation, high thermal stability, and great fire resistance, has extensive prospects for application and progress in the marine fields', aircraft, and aerospace. Nowadays, it can be extensively used as one of the advanced energy-absorbing materials in the field of sports.

The major goal of various protective equipment, clothes, and different sports mats is to safeguard the health of players and avoid sports injuries; therefore, foams used in the construction of sports equipment primarily serve a safety function. When the athlete's head collides with the sports mat first during a collision or landing, and the direction of the velocity vectors operating on the body and the head are the same, arises worst-case scenario for sports injuries.

The overall load in this scenario is concentrated in the upper body, which might result in fatal neck injuries, craniofacial injuries, and the development of localized brain damage (such as focal vascular lesions). Based on this, regardless of the sport, the most significant requirement for the foam utilized in the construction of sports equipment is superior energy absorption and impact-damping performance.

Polymeric foaming products which are designed for winter sports also perform a protective part that is akin to sports mats. This type of protective clothes has sandwich structures obtained from a hard-thermoplastic polymer with an external layer and a soft polymer foam as the internal layer. The harder layer of the thermoplastic polymers dissipates the force working on the protective environment (Tomin and Kmetty, 2021). As the foamed products are developing in the market, at the same time there are facing some major issues regarding how to dispose of the waste produced during the process? Then how to recycle? regarding the flammability characteristics and the harmful effects of BAs on the atmosphere. It is necessary to make a choroflurocarbon (CFCs) free environment by the use of more environment-friendly BAs. Researchers have to concentrate on making biodegradable foams which will progress the environmental considerations. Starch foam is an example of biodegradable foam since it was first introduced in 1989. It had been developed as a substitute for PS foams (Shastri and Stevens, 2008). Recently several researchers focus on the energy absorption characteristics of polymeric foams because of their light weight, flexibility, and facile processability. Other research areas include electromagnetic interference shielding (EMI) and strain sensor. It was reported that there is a 258% enhancement in force resistance and a 172% improvement in energy absorption regarding thermoplastic polyurethane (TPU) as it is selected for the 3D printing of the specimens (Nayak *et al.*, 2019). Presently, polydimethylsiloxanes (PDMS) serve as one of the emerging polymers or elastomers owing to their biological compatibility and outstanding stability in elasticity and chemical properties. PDMS foams are generally having great compressive resistance as well as motion energy absorption efficiency (Cai *et al.*, 2021). There are also advancements in polymeric foams as it forms multilayer foaming hierarchies through multilayer coextrusion technology, a flexible solvent-free process. By manipulating the microstructure through coextrusion technology, polymeric foams with superior properties have been observed based on the interaction, architecture of the structural levels, and scales. As a result micro and Nanolayer polymeric systems arises (Li *et al.*, 2020). Ethylene-propylene-diene monomer (EPDM) with multiwalled carbon nanotube produces highly deformable porous materials for EMI shielding. The necessity of EMI shielding ascends over the years owing to our digitally and highly linked lifestyle. EPDM/MWCNT foams are lightweight electromagnetic (EM) wave absorbers with high flexibility as well as deformability (Based *et al.*, 2020). Natural rubber (NR)-based elastomeric foams also play important roles in the foam industry. Due to elasticity, flexibility, and great strength NR, a biodegradable product has extensive commercial and strategic importance. In addition, the NR latex foam can be used as a good absorbent material, including as a chemical, sound, and oil absorbent. However, in general, rubber products cannot be used at high temperatures due to their low thermal properties. Silica is a reinforcing filler that is widely used in rubber products.

1.2 CLASSIFICATION

Polymer foams are classified regarding their cell density and average cell structure. The classification criteria are basically as follows.

- Hardness
- Density
- Foaming structure

1.2.1 HARDNESS

Depending upon the size of elastic modulus, polymeric foams are categorized as three upon 50% relative humidity and 23°C. If the value of elastic modulus, however, is less than 68.6 MPa, then polymer foam is called a soft polymer foam. If a polymer foam is said to be known as a semirigid foam, the value of elastic modulus is about 68.6–686 MPa. Rigid polymer foam arises when the value of elastic modulus is greater than 686 (Jin *et al.*, 2019). Examples include:

- Soft polyvinyl chloride (PVC) – Soft polymer foams
- Polyvinyl formal – Soft polymer foams
- Polypropylene (PP) foams – Semi-rigid
- Semi-hard PU and PI foams – Semi-rigid (Jin *et al.*, 2019).
- Hard PU – Rigid
- Phenolic foams – Rigid
- PS foams

1.2.2 DENSITY

A major criterion for classifying the type of foam is its density values. As per the values, it is reported that there are low-foaming polymeric foams, high-foaming polymeric foams, and medium-foaming polymeric foams. The value of it is less than 0.1 g/m³; such foams are considered as high-foaming and those with a density value ranging from 0.1 to 0.4 g/m³, then it is considered medium-foaming, and when a value of density is higher than 0.4 g/m³, it is low-foaming one (Jin *et al.*, 2019).

Examples: High-foaming polymers are PP foams and PI foams
Low-foaming polymers are PP and medium foaming PI

1.2.3 CELL STRUCTURE

Based on the cell structure, foams are categorized as open-cell foams and closed-cell foams. The cells are interconnected in open-cell foams and in closed-cell type; it is not connected to each other. Each cell is independent. Furthermore, they have different applications even though they are similar in their appearance. Generally, open-cell foams emerge from low-viscosity materials while closed-cell foams are more predominant in highly viscous conditions (Jin *et al.*, 2019) and (Rostami-Tapeh-esmaeil *et al.*, 2021) Open-cell foams cannot hold any gas inside the pores, but closed cell contains gas enclosed in it (Figure 1.1).

Foams made of polyurethanes are extensively utilized for open-structure foams. They have vast applications in shock absorbing, sound absorbing, and insulation purposes. These types of foams perform well because of their resisting power and dexterity toward oxygen. Apart from PU foams, NR, nitrile, EPDM, and PVC can

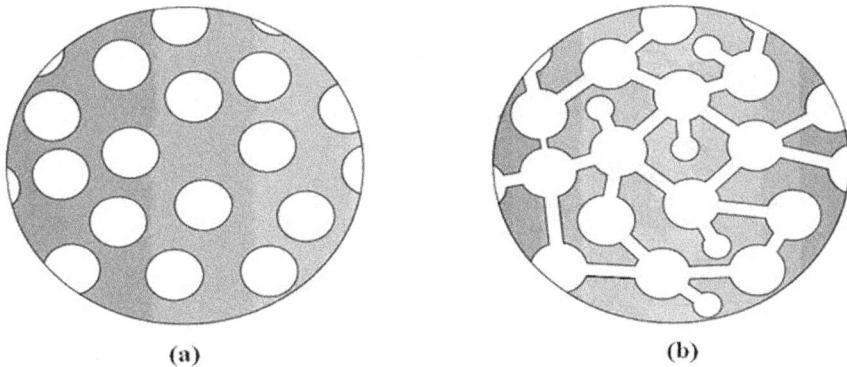

FIGURE 1.1   Illustration of polymer foam cellular structures (a) closed-cell type (b) open-cell type (Ivanova *et al.*, 2016). Each circle represents cells in the foams. Figure (a) has independent cells but figure (b) has interconnected cells.

be applied for making open cell foams. Open-cell foams generally release the gases inside the pores to the atmosphere, so that the product becomes spongier in nature as compared to closed-cell foams. Because of holes inside the matrix, open-cell foams will not have the ability to resist liquid water or water vapor. Furthermore, closed-cell foams have had pores entrapping the gases, and act as independent cells which makes them suitable for applications like packaging, automotive, construction, electronics, and marine applications. Examples include EVA, EPDM, Neoprene foam, and SBR foam.

1.2.3.1   Some Advantages and Uses of Open-Cell Foams
- More durable as compared to closed one
- The product will not shrink, break, or diminish, as time passes
- 40–50% of heat increase or decrease in apartments
- The expansion capacity of such foams is high as it is about 100 times
- Much better at soundproofing applications
- The pores in it trap most foreign particles making it good for avoiding dust particles and allergens
- Due to its special structure, it will not hold any easily vaporizable organic compounds and ozone-attacking gases
- It is attractive in reducing noise transmissions
- It is less dense as compared to closed cells ranging between 0.4 and 1.2 lbs/ft^3

Open-cell foams are extensively used for several purposes in construction, including:

- Furniture upholsteries due to its low expense
- Better applied in Interior design projects
- Much better in soundproofing
- Materials with foam protection in packaging
- Air permeability, vapor permeability, and needed application

1.2.3.2   Some Advantages and Uses of Closed Cell Foam

- Excellent in imparting heat and sound insulation
- Useful in enhancing the strength of structure
- Exterior and interior applications
- Better to reduce the transmission of vapors
- Acting as an excellent barrier to moisture
- Superior in resisting leakage

The important uses of closed-cell foams are

- The appliances and HVAC system
- Thermal insulation and shock absorption
- The seals of enclosure and cabinets
- The medical disposables
- The equipment for gas and oil
- Aerospace and aircraft
- The transportation and automotive

1.3   CLASSIFICATION BASED ON TYPE OF POLYMERS

Polymer foams are further classified into rubber and plastic, which are again divided into thermoplastic as well as thermosets-based foams. It is again divided into rigid or flexible foams. Thermoset foams have high crosslinking so that it cannot be recycled easily. Thermoplastics can be recycled (Saha *et al.*, 2005; Shastri & Stevens, 2008). In addition, thermosetting matrix generally exists in liquid forms and its cure reactions are measured through reactions showing chemical change. Some of them are phenolic resin, PU, polyester, and epoxy resin. Thermoplastic includes PE, PS, PP, and PVC. Moreover, thermoplastics melt or become soft at certain temperatures and can change into any shape. The formed shape of the polymer will not change during cooling conditions (Jin *et al.*, 2019).

1.4   BLOWING AGENTS (BAS)

BAs play a vital role in the foaming process. The properties of foams depend upon the gas production as it is created by the blowing or foaming agents, respectively. Moreover, the cell morphologies strongly depend upon the BAs. There are two types of BAs (Rostami-Tapeh-esmaeil *et al.*, 2021) (Scheme 1.1) (Tables 1.1–1.3).

Hundreds of foaming agents had been used and validated to prepare foams in the last 100 years. Nowadays, it is limited concerning some environmental issues. As per the Montreal protocol of 1987, some of the foaming agents vanished from the foaming process, viz CFCs, hydrocarbons, and liquid CO_2 owing to its ozone-depleting property (Heck, 1998). After 2003, the foaming industry completely removed the ozone-depleting-based foaming agents. The expansion of the polymer matrix will be created by the physical BAs by a change in the physical state. As a result, the gas is ejected which is to be compressed into the atmosphere on

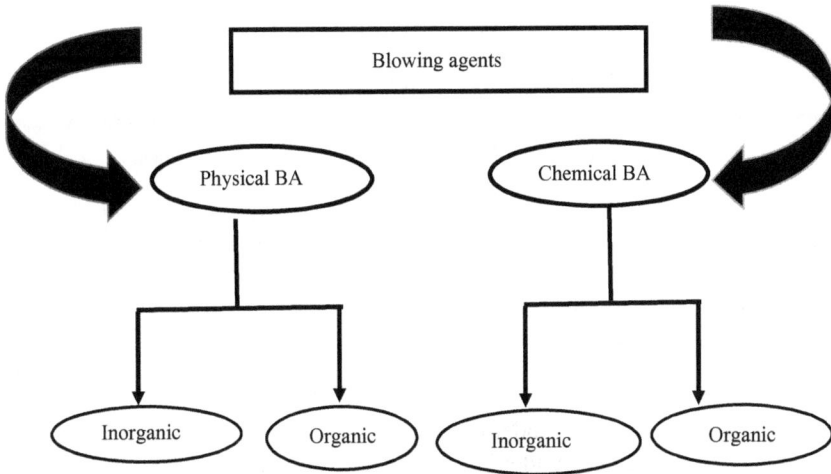

SCHEME 1.1 Showing the classification of BAs.

TABLE 1.1
Physical BAs

Inorganic	Organic
Nitrogen	Hexane
Air	Freon
Water	Hexane
Carbon dioxide	Pentane

Rostami-Tapeh-esmaeil et al. (2021).

TABLE 1.2
Chemical BAs

Inorganic	Organic
Carbonates	Para toluene sulphonyl semi carbazide (TSC)
Bicarbonates (sodium bicarbonate, ammonium bicarbonate, etc.)	ADC
	(hydrazine, vinyl, and nitrogen-based materials for thermoplastic and elastomeric foams)
Nitrates	Isocyanate/water (PU)
	5-phenyl tetrazole (5-PT)
	$4,4^1$ – Oxybis benzene sulphonyl hydrazide (OBSH)
	N, N^1 – dinitroso Penta methylene tetramine (DNPT)

Rostami-Tapeh-esmaeil et al. (2021).

TABLE 1.3
Some BAs and their decomposition temperature

Blowing agent	Gas released	Temperature in^{0}C
Isocyanate/water	Carbon dioxide	Room temperature
TSC	Nitrogen	120
ADC	Nitrogen/carbon dioxide	180
OBSH	Nitrogen	140
DNPT	Nitrogen	200

Rostami-Tapeh-esmaeil et al. (2021).

the melted polymeric matrix under pressure. CFCs, low-boiling hydrocarbons, and carbon dioxide are included under liquid physical foaming agents. Some CFCs, liquid hydrocarbons, and water serve as some examples of physical foaming agents. Physical foaming agents generally produce polyethylene and polystyrene-based non-crosslinked products with low density. It has been reported that water, carbon dioxide, CFCs produced low-density urethane foams. The production process in urethane foams introduced acetone as an auxiliary foaming agent as new development some years back. Moreover, several physical foaming agents can be used with special care in handling, storage, and process. Chemical foaming agents are a class of compounds that will produce gases or gas during chemical reactions as a result of decomposition. Furthermore, this type of foaming agent has the advantage that it exists in solid form so it has not needed any type of special storage, handling, and processing methods. Moreover, chemical foaming agents are again divided into exothermic and endothermic based on heat release and heat absorption. Generally, exothermic chemical foaming agents produce nitrogen gas and a combination of gases during a chemical reaction. As nitrogen has a slow rate of diffusion on polymer matrix, it is considered the most effective one. Carbonate-based compounds, viz sodium carbonate are the best examples of endothermic BAs as they evolve carbon dioxide on decomposition at 140°C. Being a low-temperature blowing agent, it is widely applied on rubber foams. Azo di carbonamide (ADC) is yet another foaming agent evolving 210–220 cc of gas per gram and having a decomposition temperature of about 205°C. Foaming agents are generally selected based on the decomposition and gas release temperature which has a close connection with the process temperature of the base polymer, especially for molding and extrusion methods (Heck, 1998).

1.5   PROCESS OF FOAMING

The processing steps of foaming a product mainly involve three stages

- Formation of cell
- Growth of the formed cell
- Stabilization of the cell

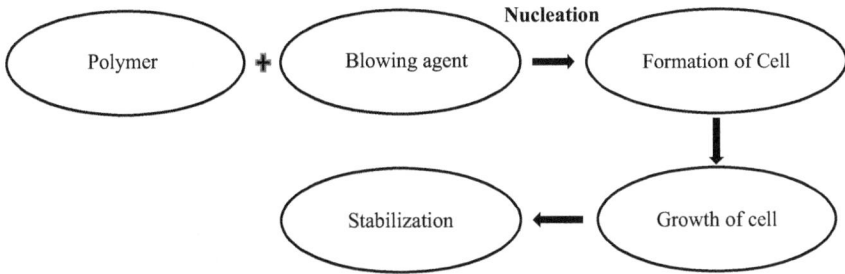

SCHEME 1.2   Showing the steps of the foaming process.

1.5.1   FORMATION OF CELL

To the molten polymer add the required quantity of foaming agents resulting in a series of chemical reactions under certain conditions yielding a large quantity of gas that will be escaped as a result of increasing the amount of gas production, which makes the solution become supersaturated. This process enables the nucleation and owing to that cell nucleus will be formed (Scheme 1.2).

1.5.2   GROWTH OF CELL

After the nucleation process, the cell begins to grow. As the size of the cell becomes smaller, the pressure will be greater inside the cavities of the cell, which means that the pressure will be inversely proportional to the radius of the cell. The cells created on the polymer matrix will not be in the same sizes and sometimes the greater pressure inside the cells resulted in the gas spreading from one small cell to another large cell means there is merging of the cell takes place. As a result, cell number increases mean nucleation, resulting in the cells with greater diameter. Thus, the growing process prevails.

1.5.3   STABILITY OF THE CELL

In order to stabilize the cell, add some surfactants or provide the cooling process to the foaming system otherwise the cells will be in an unstable condition, making the cells break and collapse. The reason is that there will be an increase in the surface area and volume of the foam system as a result of the formation and growth of cells which results in the cell walls to be thinner leading to instability.

1.6   DIFFERENT METHODS OF FOAMING

There are different foaming methods. It can be categorized as

- Chemical
- Mechanical
- Physical

1.6.1 CHEMICAL FOAMING

There are two ways to be followed. In the first method, foaming agents are added to the melted polymer and at the decomposition temperature, it will produce gas. Then apply the required amount of pressure followed by heating produce foamed product. But in the second method, inert gases are produced as a result of chemical reactions between two polymers yielding a foamed product. As it is thermally decomposing to release gas or gases in a particular temperature range, it makes the process suitable for a wide range of polymer resins that have melt viscosity in a particular temperature range. So, it can be considered a major benefit of chemical foaming. Also, this method uses a normal injection molding technique, which is taken as another important benefit. However, there are some disadvantages as well. The mold manufacturing cost is high and high precision in manufacturing the mold is essential. In some cases, like high-pressure foaming apart from the molding machine, a clamping pressure device is also needed.

1.6.2 MECHANICAL FOAMING

Through the process of mechanical stirring, the air is added to the polymer resin, making the polymer foam product. This is the vital process involved. The main advantage of using the mechanical foaming method is that it needs no blowing agent, is non-toxic with a safe and greener method of environmental protection, is low cost with high potential, is simple processing equipment, and is simple to operate. One main problem is using higher conditions on equipment.

1.6.3 PHYSICAL FOAMING

The condition of physical foaming is to use a low boiling point liquid with a polymer and it is followed by applying pressure and heat. Physical foaming agents are relatively low-cost substances, viz carbon dioxide and nitrogen. It is also a safe pollution-free method and has no residue after the foaming process. Even though it is having high potential value, it requires a special type of injection molding machine with high technical needs.

1.7 DIFFERENT PROCESSING METHODS

A wide number of processing techniques can be used for foaming. Some of the methods are given as follows (Scheme 1.3).

Foaming technologies like extrusion foaming, bead foaming, and injection molding have been accepted as the most efficient process in the industry. Nowadays, the foam market is yielding the product and facing challenges using these technologies. Nowadays, supercritical fluid technology plays an important place in the area of research and development regarding foams.

Nanocellular foams, polymer blends-based foams, semi-crystalline polymer, and copolymers-based foams, biofoams, sustainable foams, elastomeric material-based foams, Engineering plastics, and composites with high performance are recently developing and have many potential applications. In supercritical fluid technology,

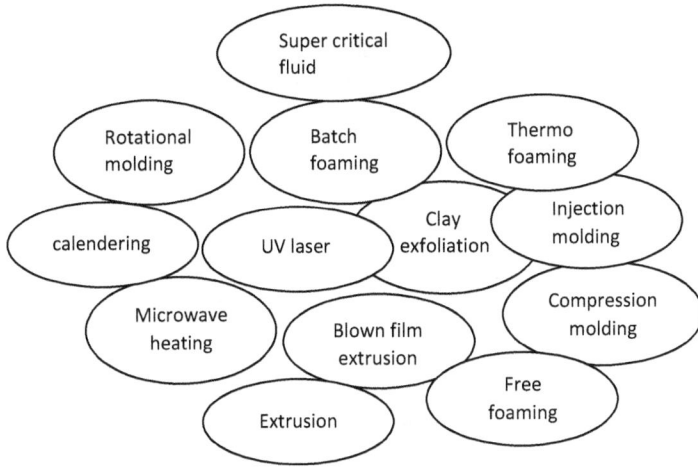

SCHEME 1.3 Showing the different processing methods.

TABLE 1.4
Showing the processing method of polymeric foams

Conditions	Foam extrusion	Batch foaming	Injection molding
Amount of sample	High amount in kilograms	Amount in grams	Amount in kilograms
State of sample	Melted form	solid	Melted form
Density of cells in cm³	10^4–10^{11}	10^6–10^{16}	10^4–10^8
Distribution of cells	Usually uniform distribution however in some cases cells in the core are having variable cell sizes from that of the edges	Distributed uniformly	Difficulty in getting uniform cell distribution
Thickness of the skin layer	Thin	Thin	Thick
Quality of the surface	Good	Glossy as well as good	Poor
Foaming agent	The foaming agent will be metered that will not be more than the melt	The material sample will be saturated with a foaming agent till an equilibrium condition is reached	The foaming agent will be metered that will be not more than the melt
Cost	Based on the machine's capacity, will be expensive	Cheap when compared with others	Mold preparation bears the extra cost

Ivanova et al. (2016).

carbon dioxide is a physical foaming agent, and a supercritical fluid creates polymer foams. There are two steps in the foaming process which are as follows:

- Formation of polymer/gas solution through the dissolution or sorption of supercritical fluid viz carbon dioxide on the polymer matrix under pressure.
- Formation of nuclei and their growth owing to a decrease in pressure
- Formation of nuclei and their growth owing to an increase in temperature (Di Maio and Kiran, 2018) (Table 1.4)

1.8   DIFFERENT FOAMS

1.8.1   THERMOPLASTIC FOAMS

A large number of thermoplastic foams are available at present. Some of them include PEN, PPSU, PEI, PES, PVC, PP, PS, and PLA. The properties of foams change as per viscosity, the morphology of cells, mechanical strength as well as melt strength (Ivanova *et al.*, 2016). Moreover, there are nanoparticle-based thermoplastic foams that also created vast advancements among thermoplastic foams. The reinforced nanoparticles enhanced the mechanical, physical, and chemical performance as well as the morphological characteristics (Ivanova *et al.*, 2016). There are current developments in polymeric foams with the new nanofillers, and most recently, the addition of nanoparticles to make the thermoplastic foams as well as all the polymeric foams having multifunctional possibilities, viz in electronics and aerospace (Antunes and Velasco, 2014). There are amorphous thermoplastic foams. The most important one is PS foams that are obtained commercially by the process of expanded bead molding, injection-molded structural foam, and expanded loose-fill packaging, extruded board, and sheet. The expanded type PS or EPS foams are fabricated from PS beads with the addition of blowing agents like pentane and some additives, followed by placing the expanded beads in a steam chest mold where it produces expansion to be needed based on density and also the bead fusion taking place. The density range for cushioning application and thermal insulation purposes is 24 kg/m^3 and 16 kg/m^3, respectively. However, these extruded PS boards are more expensive than EPS foams, but it is having a much better advantage of resisting moisture and well mechanical properties than EPS foams (Antunes and Velasco, 2014).

1.8.2   THERMOSET FOAMS

This type of polymer mainly exists in liquid form. PU, polyester, and epoxy belong to this category. Brondi *et al.*, reported that a wide variety of PU foams with high applications can be generated by applying chemical or physical foaming technologies. As a result, different modifications in the cell structures will lead to different bubble sizes and densities. This refinement causes the material to be used for particular application purposes. In general, microcellular polyurethane foams with a large number of applications are exploring in the area of research and development viz for sound absorbing, EMI shielding, thermal insulation, and other energy applications like in aeronautics, utilized as sensors in different areas due to its lightweight. Nowadays, it is also used for preparing flexible and highly sensitive low-pressure capacitive sensors for the purpose of monitoring blood pressure, interactions with human computer, detecting objects, and in the field of robotics too (Pruvost *et al.*, 2019). Generally, the preparation process involves a gelation reaction between polyol and isocyanate owing to gas evolution through mechanical, chemical, or physical processes. As the closed PU foams are having excellent thermal insulation performance due to their microcellular formation by the inert gas insertion to the pores (Di Caprio *et al.*, 2016). Ling et.al reported that epoxy foams with microcellular structures occupied a large expansion ratio through a process of supercritical carbon dioxide. The curing reaction will be initiated through higher carbon dioxide pressure. The concentration of carbon dioxide and degree of curing influenced the cell structure. In general, MCFs exhibit better mechanical performances due to a small bubble size of about 10 microns.

1.8.3 Nanocellular Foams

The main characteristics of nanofoams are based on their nanopore structure. Most of the polymeric foams had been utilized based on their micropore structures which limited their application areas. It is now a vibrant area where nanotechnology emerged with nanofillers and nanoparticles make foam products with better morphological and mechanical performances. The cell sizes are in the range of below 100 nm (Dawson) (Aram and Mehdipour-Ataei, 2016). Nanoporous foams can be prepared through selective degradation as one of the processes. In this work, one-pot method and two-step procedures can be utilized for the production of a block copolymer containing a carboxylic group which is having PS and polylactide. Supercritical carbon dioxide process is an efficient process to produce nanometric cells because it presents outstanding diffusion in the supercritical state and also is a greener one. The thermodynamic property of the polymer–gas system will sometimes produce some difficulties in the foaming process to produce nanometric cells. Basically, the nanometric cells have a large surface area and surface energies with denser nucleation of cells in the range of more than 10^{14} nuclei/cm^3. During nanocellular foaming, some particles like titanium dioxide, nanosilica, and nanoparticles will act as nucleating agents, which will enhance nucleation sites and leads to homogenous nucleation. This causes an increase in pressure drop rate and saturation pressure, causing an enhancement in nucleation rates (Notario *et al.*, 2015). A lot of studies are reporting to produce nanocellular foams. According to Costeux *et al.*, nanocellular foams having cell sizes can be obtained between 100 and 300 nm and densities in the range of 0.15 and 0.3 by using nanosilica and acrylic copolymers. By ordering the cell structure, the physical properties of nanofoams can be enhanced. When compared with conventional foams, nanofoams exhibit outstanding impact strength, tensile strength, and better toughness. When the cell size is less than 100 nm, nanofoams show better thermal insulation property due to the Knudsen effect. An ordered nanocellular structure can be created through UV induced chemical foaming process as it is a process of taking diblock copolymers with a photoacid generator followed by UV radiation (Rattanakawin *et al.*, 2020).

1.8.4 Rubber Foams

Rubber foams are generally classified as synthetic foams as well as NR foams. As these foams are having advantages when compared with thermoplastic foams, they can be utilized for several applications with regard to their strength-to-weight ratio, better flexibility, good abrasion resistance, and better energy absorption abilities. The existing applications include absorbents, thermal insulators, energy absorbers, and pressure sensors. NR latex foams were introduced in 1914 by using gas evolution chemical agents like sodium polysulphide and ammonium carbonate. However, thermoplastic foams were introduced in 1931 as it was PS. In order to produce excellent rubber foams, there are two controllable factors, viz formulation principle and processing techniques. The formulation principle includes the type and grade of the matrix and the nature of accelerating gents, filler, and foaming agents. The processing techniques include injection molding, extrusion, compression molding, and also the various conditions related to foaming like temperature, pressure, curing temperature, and precuring time. Moreover, there are open-cell and closed-cell foams with different applications (Rostami-Tapeh-esmaeil *et al.*, 2021). According to a study reported, EVA foams can be prepared using supercritical fluid carbon dioxide and analyzed its rapid expansion under various sorption temperatures resulting in open-cell and collapsed morphologies. Different studies regarding carbon dioxide in a supercritical state were investigated. Several advancements in rubber foams are explored based on nanomaterials and nanoparticles, which can act as nucleating agents in the foaming process. The higher blowing agent concentration leads to decreased elastic recovery and better resilience of blended NR/SBR foams (Rostami-Tapeh-esmaeil *et al.*, 2021). Furthermore, there are studies based on carbon dioxide adsorption from the atmosphere based on NR latex foams (Panploo *et al.*, 2019). Moreover, NR foams generally involved a sulphur vulcanization process that will impart better physical properties for the product for a particular application. Even though there are alternative methods for vulcanization like radiation and peroxide usage, the sulphur system gained acceptance in the industry (Pornprasit and Aiemrum, 2018). But peroxide vulcanization can be regarded as environmentally friendly due to the absence of non-toxic heavy metal ions. Recently research developments are continuing based on NR foams as it is biodegradable when compared with other petroleum-based products (Ismail and Ariff, 2016). There are two different pathways for the preparation of NR sponges as well as foams (Service and Wildlife, 1986).

1.8.5 BIOFOAMS

Apart from other foams, there are biofoams that also play a vital role in many applications as well as environmental considerations. Different studies, as well as different pathways, are still in progress for producing biodegradable foamed materials like jute, wood- flour, hemp, and starch with degradable polymers and also use cellulose, starch, polycaprolactone, aliphatic polyesters, and aromatic polyesters (Khemani, 1997). It is a great concern for the foaming industry to reduce petroleum-based foaming products and make a substitute for several applications. Taking this into consideration, much more reports or studies are still concentrating to produce biofoams from biocompatible materials. Biopolymers are used for preparing biofoams. Most of the studies are based on polylactic acid and starch and still, studies regarding cellulose-based biopolymers are available. Cellulose-based derivatives are generally originating from plants like fibers of leaves, and cotton. Cellulose particle foams are to be created by a twin screw extruder. The process involves the mixing of all the ingredients with the polymer and isopentane can be used as a blowing agent for the preparation of foam. The melt temperature was set to be constant and the pressure will be varied. There is a relation connecting material characteristics and the foaming of biopolymers. During this process, the cell size is reduced to below 100 µm (Rapp *et al.*, 2014). Recently there are studies on cellulose-based nanocrystal foams with exceptional thermal insulating properties and here polybutylene succinate was the precursor used. This type of foam can be applied to construction, packaging, tissue engineering, etc. (Yin *et al.*, 2020). Nanocomposite foams reinforced with nanocellulose are also emerging studies for utilizing various applications. Apart from cellulose, there are starch foams that can be applied in disposable-type packaging of materials. The concentration of amylose content in the starch will give rise to foams with better strength and density. Moreover, high amylopectin produced lighter foams with the least strength (Soykeabkaew *et al.*, 2015).

1.8.6 OTHER POROUS FOAMS

There are some other materials that can be utilized for foam preparations. They are concrete, metals, and ceramics. There are also elastomeric-based liquid metal (LM) foams that can be used for tactile sensing. It is an ultra-soft composite material with high or low permittivity in stimulus to a particular compression. They have wide applications as soft sensors and soft capacitors and as a result, they can be applied to smart robotics, health care monitoring, and electronic skins. The preparation procedure includes a LM viz eutectic gallium indium with sugar particles which will create porosity on LM elastomer (Yang *et al.*, 2020). There are also some other studies regarding stretchable-type LM elastomer for the purpose of controlling EM shielding and in the field of flexible electronics (Yu *et al.*, 2020). Another study reported producing of foams based on LM elastomer for the purpose of force sensing and triboelectric energy harvesting (Nayak *et al.*, 2019). There are also ceramic foams-related studies from 2001 to 2016 as this type of foam has also potential applications in filtering, scaffolds in the field of tissue engineering, catalytic reaction that supports energy harvesting. Ceramic foams can be prepared by different processing methods like dip coating and thermal treatment methods (Fey *et al.*, 2017). Concrete foams are yet another porous material that can be reinforced by inserting nanoparticles and an organic surfactant. Finer pore structures are developing through the inclusion of nanoparticles (She *et al.*, 2018) and organic surfactants. Another category of porous materials includes aerogels a type of foam material which are explored for several applications. Porous structures in aerogels are obtained by removing the liquid solvent of a gel.

1.9 CONCLUSION

Polymeric materials are lightweight porous materials with enormous applications like cushioning, packaging, automotive, energy absorption, thermal insulation, sound absorption, shock absorption, flexible electronics, water purification, and oil absorption. There are thermoplastic-based as well as elastomeric-based foams and their applications are based on the pore density, pore structure, and the type of cells formed. Open-cells, as well as closed-cell foams, are having a different types of applications based on the nature of the cell connections. Polymeric-based foams are generally prepared through injection molding, extrusion, supercritical fluid technology, and batch foaming. The growing use of foam-based products makes the processing methods to be more innovative and successful. This leads to various microcellular foams. As time passes, nanotechnology and scientific developments opened a new way of enhancing or modifying MCFs with the addition of nanofillers or nanoparticles. This leads to the formation of nanofoams with outstanding surface and morphological properties. Moreover, some other porous materials other than polymeric foams are available with extensive applications.

REFERENCES

Antunes, M. and Velasco, J. I. (2014) 'Multifunctional polymer foams with carbon nanoparticles', *Progress in Polymer Science*, 39(3), pp. 486–509. doi: 10.1016/j. progpolymsci.2013.11.002.

Aram, E. and Mehdipour-Ataei, S. (2016) 'A review on the micro- and nanoporous polymeric foams: Preparation and properties', *International Journal of Polymeric Materials and Polymeric Biomaterials*, 65(7), pp. 358–375. doi: 10.1080/00914037.2015.1129948.

Based, A. *et al.* (2020) 'Highly deformable porous electromagnetic wave nanotube nanocomposites', *Polymers*, 12(4), p. 858.

Cai, J. H. *et al.* (2021) 'Thermo-expandable microspheres strengthened polydimethylsiloxane foam with unique softening behavior and high-efficient energy absorption', *Applied Surface Science*, 540(P1), p. 148364. doi: 10.1016/j.apsusc.2020.148364.

Di Caprio, M. R. *et al.* (2016) 'Polyether polyol/CO2 solutions: Solubility, mutual diffusivity, specific volume and interfacial tension by coupled gravimetry-axisymmetric drop shape analysis', *Fluid Phase Equilibria*, 425, pp. 342–350. doi: 10.1016/j.fluid.2016.06.023.

Fey, T. *et al.* (2017) 'Reticulated replica ceramic foams: Processing, functionalization, and characterization', *Advanced Engineering Materials*, 19(10), pp. 1–15. doi: 10.1002/adem.201700369.

Heck, R. L. (1998) 'A review of commercially used chemical foaming agents for thermoplastic foams', *Journal of Vinyl and Additive Technology*, 4(2), pp. 113–116. doi: 10.1002/vnl.10027.

Ismail, H. and Ariff, Z. M. (2016) 'Rubber Latex Foam', *Polymer-Plastics Technology and Engineering* 51(15), pp. 1524-1529. doi: 10.15376/biores.11.1.1080-1091.

Ivanova, N. *et al.* (2016) 'We are IntechOpen, the world's leading publisher of Open Access books Built by scientists, for scientists TOP 1%', *Intech*, i(tourism), p. 13.

Jin, F. L. *et al.* (2019) 'Recent trends of foaming in polymer processing: A review', *Polymers*, 11(6), p. 21. doi: 10.3390/polym11060953.

Khemani, K. C. (1997) 'Polymeric foams: An overview', *ACS Symposium Series*, 669, pp. 1–7. doi: 10.1021/bk-1997-0669.ch001.

Li, Z., Olah, A. and Baer, E. (2020) 'Micro- and nano-layered processing of new polymeric systems', *Progress in Polymer Science*, 102, p. 101210. doi: 10.1016/j.progpolymsci.2020.101210.

Di Maio, E. and Kiran, E. (2018) 'Foaming of polymers with supercritical fluids and perspectives on the current knowledge gaps and challenges', *Journal of Supercritical Fluids*, 134, pp. 157–166. doi: 10.1016/j.supflu.2017.11.013.

Nayak, S. *et al.* (2019) 'Liquid-metal-elastomer foam for moldable multi-functional triboelectric energy harvesting and force sensing', *Nano Energy*, 64(May), p. 103912. doi: 10.1016/j.nanoen.2019.103912.

Notario, B. *et al.* (2015) 'Experimental validation of the Knudsen effect in nanocellular polymeric foams', *Polymer*, 56, pp. 57–67. doi: 10.1016/j.polymer.2014.10.006.

Panploo, K., Chalermsinsuwan, B. and Poompradub, S. (2019) 'Natural rubber latex foam with particulate fillers for carbon dioxide adsorption and regeneration', *RSC Advances*, 68, pp. 28916–28923. doi: 10.1039/c9ra06000f.

Pornprasit, P. and Aiemrum, A. (2018) 'Natural Rubber Latex Foam for Seedling', *10th Int Conf Sci Technology Sustainable Well-Being (STISWB 2018)*, pp. 515–519.

Pruvost, M. *et al.* (2019) 'Polymeric foams for flexible and highly sensitive low-pressure capacitive sensors', *NPJ Flexible Electronics*, 3(1), pp. 13–18. doi: 10.1038/s41528-019-0052-6.

Rapp, F., Schneider, A. and Elsner, P. (2014) 'Biopolymer foams - relationship between material characteristics and foaming behavior of cellulose based foams', *AIP Conference Proceedings*, 1593(February 2015), pp. 362–366. doi: 10.1063/1.4873801.

Rattanakawin, P. *et al.* (2020) 'Highly ordered nanocellular polymeric foams generated by UV-Induced chemical foaming', *ACS Macro Letters*, 9(10), pp. 1433–1438. doi: 10.1021/acsmacrolett.0c00475.

Rostami-Tapeh-esmaeil, E. *et al.* (2021) 'Chemistry, processing, properties, and applications of rubber foams', *Polymers*, 13(10), pp. 1–53. doi: 10.3390/polym13101565.

Saha, M. C. *et al.* (2005) 'Effect of density, microstructure, and strain rate on compression behavior of polymeric foams', *Materials Science and Engineering A*, 406(1–2), pp. 328–336. doi: 10.1016/j.msea.2005.07.006.

Service, A. N. P. and Wildlife (1986) 'Special issue. Special issue', *Australian Ranger Bulletin*, 4(1), pp. 9–10.

Shastri, V. and Stevens, M. (2008) 'Polymer Foams', *Encyclopedia of Biomaterials and Biomedical Engineering, Second Edition - Four Volume Set*, pp. 2270–2274. doi: 10.1201/b18990-217.

She, W. *et al.* (2018) 'Application of organic- and nanoparticle-modified foams in foamed concrete: Reinforcement and stabilization mechanisms', *Cement and Concrete Research*, 106, pp. 12–22. doi: 10.1016/j.cemconres.2018.01.020.

Soykeabkaew, N., Thanomsilp, C. and Suwantong, O. (2015) *A review: Starch-based composite foams, Composites Part A: Applied Science and Manufacturing*. Elsevier Ltd. doi: 10.1016/j.compositesa.2015.08.014.

Tomin, M. and Kmetty, Á. (2021) 'Polymer foams as advanced energy absorbing materials for sports applications—a review', *Journal of Applied Polymer Science*, 2021, pp. 1–23. doi: 10.1002/app.51714.

Yang, J. *et al.* (2020) 'Ultrasoft liquid metal elastomer foams with positive and negative piezopermittivity for tactile sensing', *Advanced Functional Materials*, 30(36), pp. 1–10. doi: 10.1002/adfm.202002611.

Yin, D. *et al.* (2020) 'Fabrication of branching poly (butylene succinate)/cellulose nanocrystal foams with exceptional thermal insulation', *Carbohydrate Polymers*, 247(June), p. 116708. doi: 10.1016/j.carbpol.2020.116708.

Yu, D. *et al.* (2020) 'A super-stretchable liquid metal foamed elastomer for tunable control of electromagnetic waves and thermal transport', *Advanced Science*, 7(12), pp. 1–9. doi: 10.1002/advs.202000177.

2 Micromechanical Modelling of Foams

Anu Sasi, Jobin Jose, Pradeep Bhaskar,
Ratheesh P.M. and Jinu Jacob George

CONTENTS

2.1 INTRODUCTION

Polymeric foams have wide areas of applications because of their properties like good strength-to-weight ratio, high insulation to heat and sound, damping properties, lightweight, and resistance to impact. The major industries where polymer foams are currently used are automotive, sports, footwear, marine, packaging, electronics, aerospace, building and construction, mattresses, and medical applications [1]. The important foaming methods are chemical, mechanical, and physical. During the manufacturing process, the gas bubbles nucleate and grow in the melted or softened polymer [2]. After the production process, cellular solids having cell sizes in various ranges will be obtained. The foam produced has cells arranged in a three-dimensional manner and which will fill the spaces, comprised of the membranes, struts, and concentrations of the polymer at the joining vertex of struts [3].

Polymer foams can be prepared generally by using various processing techniques of foaming such as batch foaming, bead foaming, injection foaming, and extrusion foaming [4,5]. The foams can be classified from nanoscale to microscale according to the size of cells. Foams having cell sizes in the range of 0.1–100 nanometres

DOI: 10.1201/9781003218692-2

are considered nano-size foams. Cell sizes in the range of 0.1–1 μm are called ultra-micron-sized foams, 1–100 μm sized cell foams are named micron-sized foams, and foams having large cell sizes are known as macro-sized foams [6]. The major factors that control the mechanical properties of foams are the properties of the polymer used, the geometry of the cell, different manufacturing methods, and the pressure of the gas (closed-cell foams) [7].

Since the 1930s, various thermoplastic and thermoset polymers are used for the manufacturing of polymeric foams. They are polystyrene, ethylene propylene diene monomer, poly(methyl methacrylate), polyurethane (PU), polylactic acid, poly(caprolactone), polyetherimide, polyethylene, poly(vinyl chloride), polycarbonate, polypropylene (PP), and poly(vinyl chloride) [6]. There are various types of cell microstructures available but it is difficult to manufacture the foams with a specific microstructure. Based on the constituent materials and microstructure, one can predict the mechanical properties of the foam. Such prediction of mechanical properties is called micromechanical modelling [8].

The major parameters that rely upon the micromechanical properties of foams are [3,9–13].

1. Cell geometry of foam
2. Type of polymer used
3. Gas pressure
4. Relative foam density

2.1.1 Cell Geometry of Foam

The geometry of foams is an important parameter and will determine the behaviour of foams. The foam consists of a network of space-filling polyhedra. The wall surrounding the cells are called faces. The borders of faces are known as struts. The vertices where the struts intersect are named knots [14].

According to Euler's law, for any foam structure, the number of cells (C), edges (E), faces (F), and vertices (V) of cells in the 3D was related by Equation 2.1

$$-C + F - E + V = 1 \qquad (2.1)$$

The geometrical features of the foam structure like cell size, shape, or geometrical anisotropy, the relative density of the foam, the relative number of cells in the foam, the thickness of cell wall, and the thickness of polymer between the struts and faces are the important constituents.

(a) Relative number of open cells

Polymer foams can be either open or closed cell structures based on their cell geometry (Figure 2.1).

In the open cell foams, there was a series of interconnecting cells with open structures, which is the major difference compared to closed cell structures. But in practice, a foam has both closed and open cells. Hence the

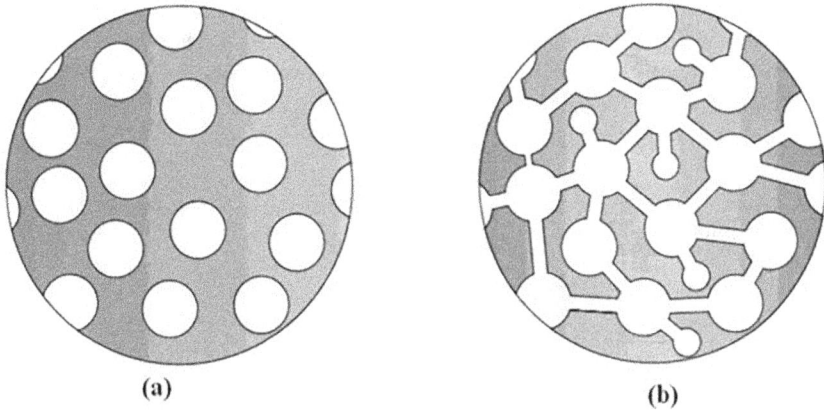

FIGURE 2.1 Cellular structure of (a) closed and (b) open cell polymer foam [1]. (open access, No copyright required).

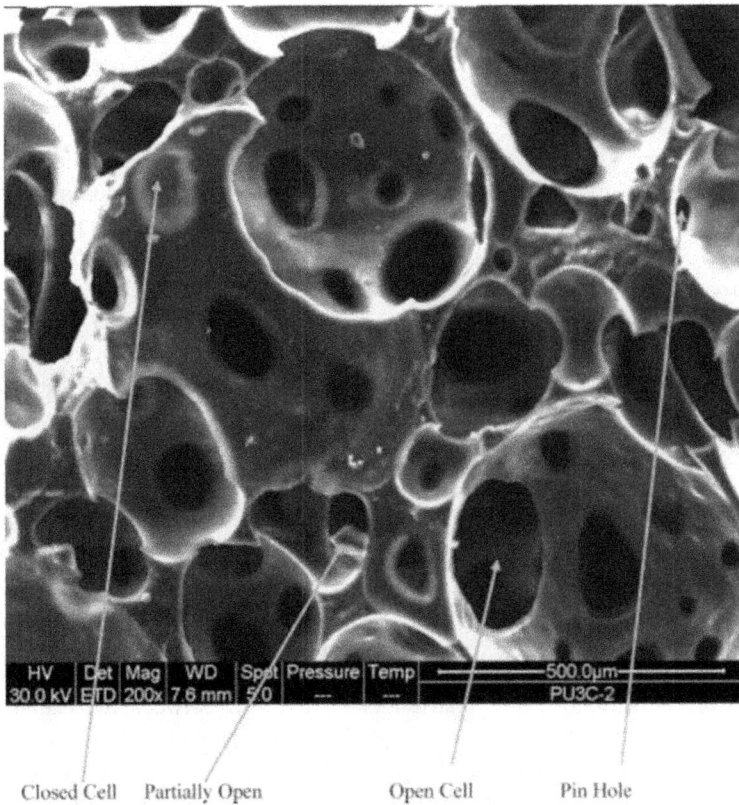

FIGURE 2.2 Cell window flexible PU foam [15]. (with permission from ACS).

properties of cellular material depend on the degree of closed or open cells. In the closed cell, the relative number of open cells is related to the relative density of the foam, also depends on the physical properties of cellular material. ESEM image of a flexible PU foam with different cell windows such as closed, pinholes, partially open and open is shown in the figure below [15].

The effective number of open cells in a foam, P_{eff} can be calculated using Equation 2.2.

$$P_{effective} = \frac{N_{open} + \left(\frac{1}{2}\right)N_{part.}}{N_{open} + N_{part.} + N_{pin} + N_{closed}} \qquad (2.2)$$

N is the number of walls of closed, pinholes, partially open and open cells.

In practice, it is difficult to manufacture foams having only closed-cells or open-cells structures. The open-cell foams have >90% of cells open, similarly, most of the cells in a closed-cell foam are closed [3].

(b) **Cell size**

Cell size is an important parameter to consider during the determination of foam properties. Because of the difference in the cell size distribution of foam, during compression, the cells will not collapse uniformly. Foams having uniform cell size will collapse uniformly with the application of compressive force. But in some foams the initial loading causes the collapse of outside layers, the cells in the centre portion will distort very little [7,16,17]. Thus, the mechanical behaviour of foams depends on both the cell size and also the distribution of cell sizes within the foam [3].

(c) **Cell shape**

The final structure of polymeric foams is mostly isotropic due to the difference in the manufacturing process. Hence the properties of foams are dependent on their production process. Due to different cell dimensions in different directions, commonly three main directions are determined in the foams. In the directions having higher foam dimensions, the mechanical characteristics of foams are found to be better [3,7].

(d) **The thickness of cell walls and distribution of polymer between the walls and struts**

In closed-cell foams, the polymer material is mostly concentrated in the walls, struts and vertices. But in open-cell foams, it is concentrated mainly in struts. The polymer material distributed in the vertices of low-density foams is negligible. The distribution of polymer between the walls and struts has a direct influence on the foam's mechanical properties [3].

(e) **Foam cell geometry and its constituents**

To predict the mechanical properties of foam, structural elements and properties of the solid material used should be thoroughly studied. The cell geometrical parameters in foam are cell wall, strut etc. According to

Plateau law[18], intersection of a minimum of three walls is called a strut. An intersection of four struts or an intersection of six walls is called a knot point.

The thickness of the cell wall is not uniform over the entire part of foam cells. There is a gradual decrease in the cell wall thickness from the wall border to the centre. The struts are thicker at the vertices and thinner at the centre part. The cross-section of the struts and the distribution of wall thicknesses are also varied across different portions of the foam. Therefore, considering the mean values of these parameters in the modelling will cause severe errors. Hence, the measure of walls thickness, struts cross-sections and distribution of these over the entire foam is important. Various models which deal with the variation of wall thickness and strut cross-sections are important for the micromechanical modelling of foams [3,7].

2.1.2   RELATIVE FOAM DENSITY

The relative density of the foam is an important factor that determines the various physical properties of the polymeric foams. Foam density is calculated by dividing the mass of the foam by the volume of foam. Since the foam structure is not homogeneous there will be a density difference between the surface layers and average foam density. Hence we consider relative foam density to express the density distribution.

The mechanical properties of polymeric foams can be connected to the relative density of foam by the given formula.

$$\frac{\varphi_f}{\varphi_s} = C\left(\frac{\rho f}{\rho s}\right)^p \tag{2.3}$$

where φ_f and φ_s are the mechanical properties of foam and solid phase, respectively.

For a polymeric foam, the relative density is directly related to the dimensions of foam cells by the following equation [19]

$$\frac{\rho_f}{\rho_s} \alpha \frac{V_f}{V_s} \alpha \left(\frac{t}{l}\right) \tag{2.4}$$

For a closed-cell foam, the stiffness is related to relative density by Equation 2.5 be attributed to these three contributions [20]

$$\frac{E^*}{E_s} \approx \varphi^2 \left(\frac{\rho^*}{\rho_s}\right)^2 + (1-\varphi)\frac{\rho^*}{\rho_s} + \left(\frac{p_0}{E_s}\right)\frac{(1-2\vartheta)}{1-\frac{\rho^*}{\rho_s}} \tag{2.5}$$

The modulus of the foam will change directly with the square of the foam density. When $\varphi = 1$, the above equation can be used for the determination of foam stiffness having an open cell structure.

2.1.3 GAS PRESSURE

The cavities of a closed-cell foam are filled with gaseous material and the presence of a gaseous substance will lead to the formation of internal gas pressure on the cell walls or foam matrix. Gibson & Ashby [20] modelled a closed-cell foam having isotropic symmetry and they provide a relation between Young's modulus of polymer used, Young's modulus of foam cell in terms of the fraction 'φ' of polymer in the struts as in Equation 2.5, where P_0 is the internal gas pressure of the cells.

In synthetic foams, the pressure of the gas inside the cells is nearly equal to the atmospheric pressure (0.1 MPa), hence the last term in Equation 2.5 can be neglected.

Stress during the plastic collapse Σ_{Pl} is related to the yield stress of the polymer material $\sigma_{y,p}$ under the compression, loading is given by

$$\frac{\Sigma_{Pl}}{\sigma_{y,p}} = C_1 \left(\phi \frac{\rho_f}{\rho_s} \right)^{3/2} + C_1' \left(1 - \phi \right) \frac{\rho_f}{\rho_s} + \frac{\left(p_0 - p_{at} \right)}{\sigma_{y,p}} \tag{2.6}$$

C_1 and C_1' are constants and are equal to 0.3 and 1, respectively. These are obtained by fitting the experimental data [3,20].

Sun et al. and Oscar Lopez-Pamies et al. and investigated the effect of internal gas pressure on the mechanical properties of 2D cellular structures [21] and elastomeric closed-cell foams [22], respectively.

2.1.4 PROPERTIES OF POLYMER MATERIAL

The major features of foams depend on various properties of the polymer material used, cell geometry, the manufacturing method in which the foam is made and also gas inside the foam (closed-cell) foam. The mechanical properties of the polymer used will change after solidification and the foaming process of foam manufacturing. During foaming due to the alignment of the polymer molecules in the orientation direction of materials, the mechanical properties are better in the direction of orientation. Hence, it is necessary to calculate the properties of foam solid material before and after the foaming process.

The solid material inside the foam is considered based on the final mechanical properties required during the service life of the foam. If the required properties are concerned only with the linear properties of the foam, then analysing Young's modulus only may be sufficient. For advanced models, various properties of the solid materials are considered. Fracture of foam is also considered in the model. Since the polymer is a time-dependent material, this is also incorporated into the model. The foams are very sensitive to strain rate during the loading condition and for analysing the cushion performances static tests are also taken into consideration [7,23,24].

2.2 MODELLING OF POLYMERIC FOAMS

The work of Jeong et al. investigates the dependence of the rate of strain on the behaviour of PU foam and introduces a new model at various strain rates to improve the experimental data. At two different strain rates, the quasi-static compression tests of these foams are carried out and depend on the seven parameters [25].

2.2.1 QUASI-STATIC MODELS

The stress-strain curve of the polymeric foam can be divided into three sections an elastic region of a linear shape, a plateau region having plasticity properties and a final densification region where the stresses rise suddenly as shown in Figure 2.3 [26].

Three equations that describe different regions in the polymeric foam stress-strain curve are shown below:

Linear elastic region:

$$\sigma = E\varepsilon \text{ if } \sigma \leq \sigma_{yield} \tag{2.7}$$

Plateau region:

$$\sigma = \sigma_{yield} \text{if } \varepsilon_{yield} \leq \varepsilon \leq \varepsilon_D \left(1 - D^{-\frac{1}{m}}\right) \tag{2.8}$$

Densification region:

$$\sigma = \sigma_{yield} \frac{1}{D} \left(\frac{\varepsilon_D}{\varepsilon_D - \varepsilon}\right)^m \text{ if } \varepsilon > \varepsilon_D \left(1 - D^{-\frac{1}{m}}\right) \tag{2.9}$$

where σ – engineering stress, ε – engineering strain. The other important parameters of this model are E Young's modulus at the elastics portion, σ_y is yield stress, ε_D is the strain at the densification region, and D and m are two constants. In this

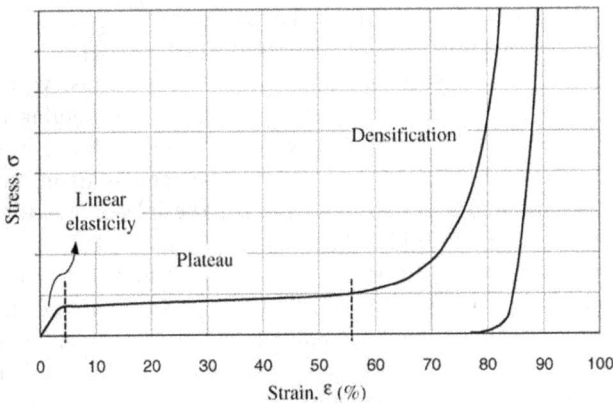

FIGURE 2.3   Stress-strain curve of PU foam (quasi-static) [29].

model, stress is almost constant in the plateau region, but at the two boundaries, the stress-strain graph has a sudden change [13,27,28].

The Rusch model is phenomenological, not a micromechanical model and is the sum of two power laws:

$$\sigma = A\varepsilon^m + B\ \varepsilon^n \tag{2.10}$$

where the four important parameters A, B, m and n, can be determined empirically. σ is the engineering stress, ε is the engineering strain and has a positive magnitude during compression. This model can explain the stress–strain relation by only one equation, but this model is accurate for describing the densification phase of the foam [27,29,30].

Avalle et al. discussed a model to fit the experimental stress-strain values of the foam material which is shown in the below equation. The parameters such as B, A and E directly depend on the density, but m and n are not [27].

$$\sigma(\varepsilon) = A\left(1 - e^{-(E/A)\ \varepsilon(1-\varepsilon)^m}\right) + B\left(\frac{\varepsilon}{1-\varepsilon}\right)^n \tag{2.11}$$

Jeong et al. discussed two different models for both the high and low and density foams based on PU. At various strain rates, this model can be used to determine the stress strain of the discussed foams. Stress-strain data at a different rates of strain are determined by dynamic compression tests and the result obtained was compared with the constitutive model [25].

In current years the importance of Finite element analysis (FEA) is used for solving the micromechanical modelling of cellular materials. FEA was used extensively to solve the micromechanical models for fracture toughness estimation of cellular materials. The foams are made up of non-linear material that is arranged in the form of a complex structure. Depending on the application, the polymer foams are subjected to different types of deformations. There are various finite element techniques available to solve these micromechanical problems. FEA is a method of standard industrial test which is used to identify the large deformation behaviour of foams, by finding out foam compression between two parallel plates [7,31].

To investigate the microstructure and properties of foams, various types of foam micro models are available and can be classified into regular and irregular models based on the distribution of cells within the foams. Since it is difficult to manufacture foams with 100% regular lattice, the regular foam models are an approximation of the real foam structure. While the irregular is the representation of the real foam structure [32]. For both the closed and open-cell foams the micromechanics is different. In open and closed cell foams the most important regular models are proposed by Gibson and Ashby and Kelvin model. Foams with open-cell irregular structure can be explained by two important models and they are Zhu's Voronoi tessellation Model and Vander and Burg Model. For closed-cell foam, the irregular models such as the Roberts–Garboczi model and Shulmeister model models are the most important ones as shown in Figure 2.4 [3,7,31]. Yu et al. proposed a unit cube geometrical model for the characterisation of the

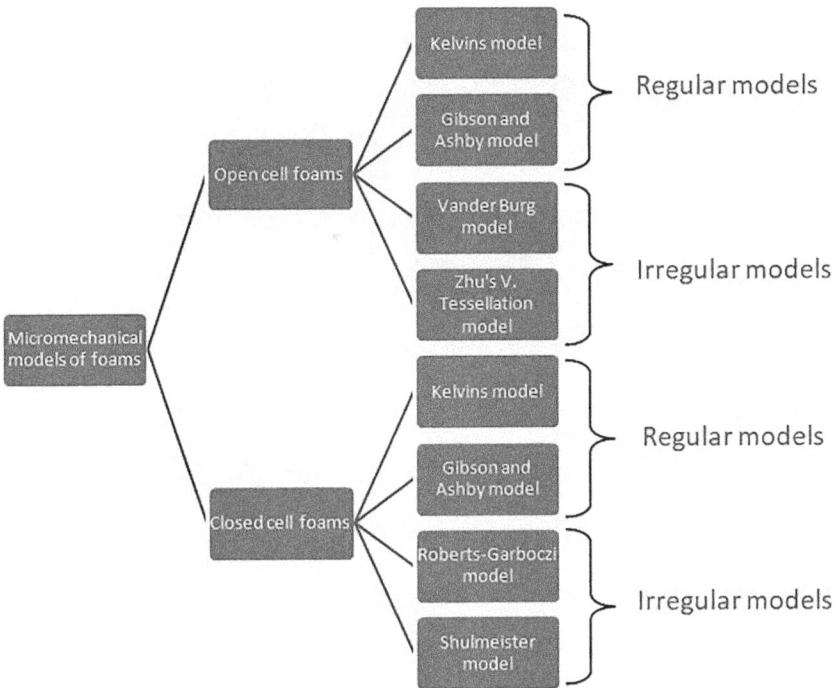

FIGURE 2.4   Representation of available foam models [7].

internal structure of porous carbon foam. The geometrical parameters required for the calculation of transfer and flow of heat through the porous foam can be derived using this model [33].

2.2.2   Open-Cell Foams and Their Applicability

Open-cell foams also called flexible foams, have the morphology of connected cell morphology. These open-cell foams have good absorption properties when compared with closed-cell foams. The cell walls of the foams having open cell morphology are in a state of fractured or disconnected. Since the cell walls of the open-cell structure are in the broken stage and these structures were dominated by the ribs and struts. The most important applications of flexible foams are cushioning sound absorption and insulation, vibration dampening and temporary frameworks for the development and binding of cells in tissue engineering, etc. [6,7]. Figure 2.5 shows a diagram of an open-cell structure, where the most of cells are interconnected in nature. For open-cell foam, the struts are interconnected only at the joints.

If a foam having an open-cell structure is loaded, the cell topology causes the cell edges to bend. If the cells are closed then the cell faces can buckle easily but here also the obtained deformation will be mostly bending [35,36].

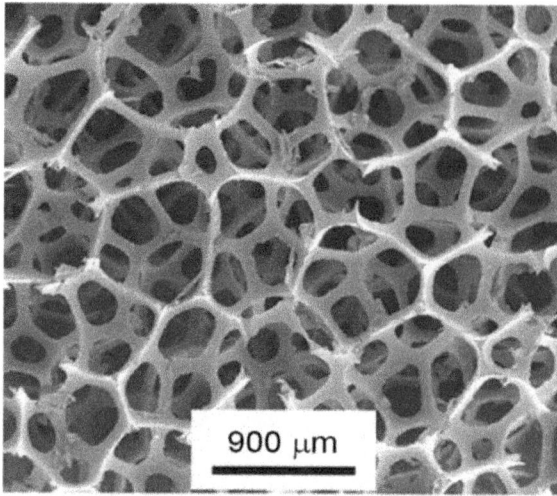

FIGURE 2.5   A typical open cell structure [34].

2.2.3   REGULAR MODELS

The regular foam models are the ideal foam models where the foam cell shapes are assumed to be space-filling polyhedrons [37].

These types of models simplify the foam cells as a simple polyhedron.

The most commonly used space-filling polyhedrons are the cubic model, Kelvin model and Weaire–Phelan model.

Gibson and Ashby model: the Ashby and Gibson model explains different mechanical properties of 3D and 2D cells [20,38]. The mechanical properties of 2D cellular materials in both nonlinear and linear plastic and elastic materials or honeycombs are calculated and compared with experimental results. These properties of cellular materials can be explained in various conditions such as elastic buckling, bending, plastic collapse etc. [13].

A simple model of open-cell foams was given by M. F. Ash and L. J. Gibson and depicted the foam as a cubic cell arrangement having length, 'l' and struts thickness t as in Figure 2.7 [19].

The volume of material in the strut V_{srtut} can be calculated using Equation 2.12.

$$V_{strut} = t^2 l \tag{2.12}$$

The struts are shared by three adjoining cells, in each cell, the relative volume of solid material can be determined by

$$V^* \geq \frac{12}{3} t^2 l \tag{2.13}$$

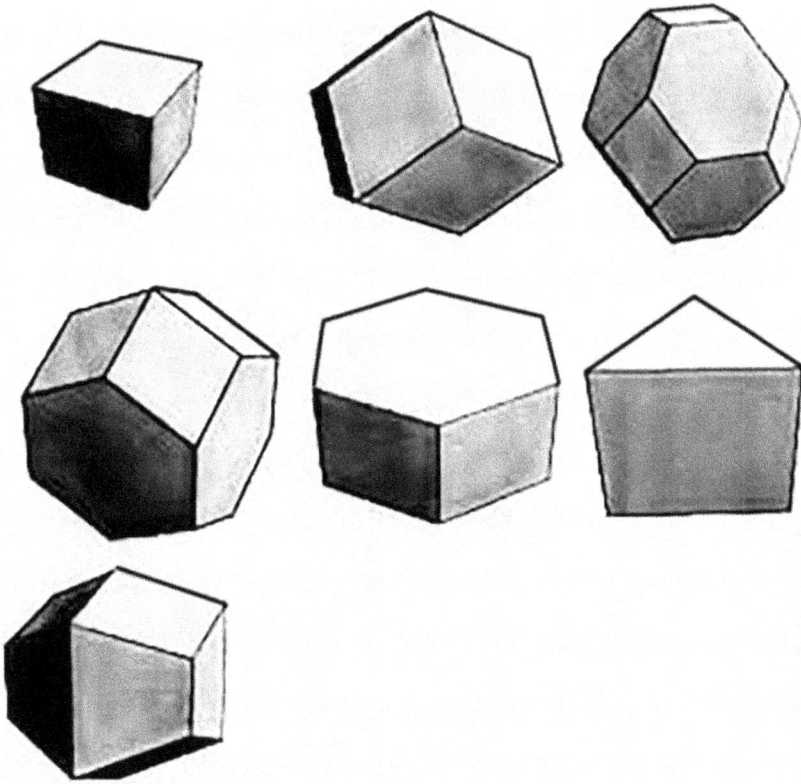

FIGURE 2.6 Space-filling polyhedron.

The relative foam density of the structure of foam can be related to the dimension of the cell by the equation,

$$\frac{\rho^*}{\rho_s} \propto \frac{V^*}{V_c} \propto \left(\frac{t}{1}\right)^2 \tag{2.14}$$

The V_c is cell volume, 1^3

The elastic modulus of foam can be calculated from the elastic deflection of a beam with a length 1 which is loaded by a load at the midpoint.

The Beam theory can explain the deflection, δ in the figure using the below equation [39,40]

$$\delta \propto \frac{F1^3}{E_s I} \tag{2.15}$$

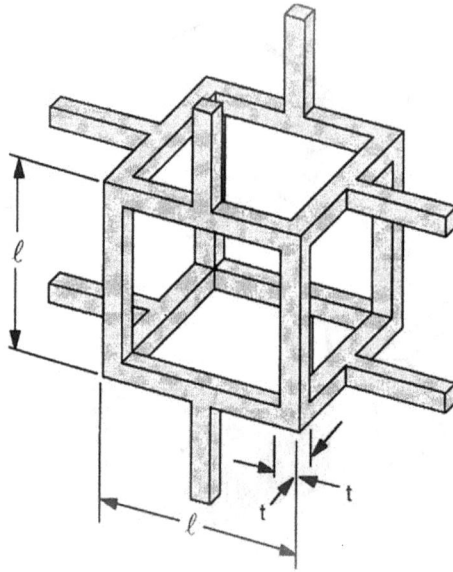

FIGURE 2.7 Open cell unit cell having cubic symmetry. The cell is composed of edges with length l and thickness of struts 't' [19].

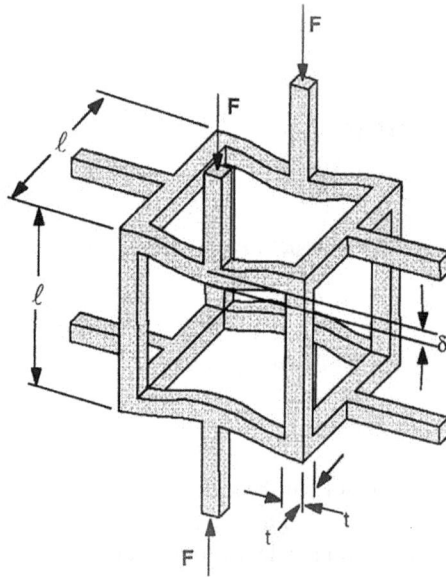

FIGURE 2.8 After linear elastic deflection of a unit cell with a magnitude of deflection δ induced by a force F [19].

where 'I' is the moment of inertia for a strut that is point-loaded. If the foam is subjected to uniaxial remote stress, the magnitude of the force transmitted on each strut is equal to F, then this structure will undergo deflection as per the above equation.

The modulus of the foam can be determined using the below equation

$$E^* = \frac{\sigma}{\varepsilon} = \frac{C_1 E_s I}{l^4} \tag{2.16}$$

C_1 is the geometrical parameter of foam, which is obtained by fitting the above equation with the obtained data [19].

The Gibson and Ashby models showed that C_1 is nearly equal to 1 for most materials [19,20,38]. The below equation explains the relation between density and elastic modulus for open-cell forms having very small displacements [19].

$$\frac{E^*}{E_s} = \left(\frac{t^4}{l^4}\right) = \left(\frac{\rho^*}{\rho_s}\right)^2 \tag{2.17}$$

This equation assumes that all the polymer material in foam is concentrated at the region of struts, hence only the thickness of the struts is taken into consideration. But in the closed-cell foam, it is different and is discussed in Section 3.2 [19].

Kelvin model is the space-filling orderly arrangement of cells having equal volume. The polyhedron used for the space-filling of a Kelvin foam is called tetrakaidecahedron cells which have 14 faces and 24 vertices. The Kelvin foam formed from these arrangements has a low surface area. William Thomas in his work explains the micromechanics of both open and closed-cell foams, explained using the Kelvin model [41–44].

Kucherov et al. explained the brittle fracture behaviour of open kelvin foam. In this type of model, the foam is considered a spatial lattice, which has brittle-elastic struts connected at nodal points [45].

Zhu et al. described that for open-cell foam models which have tetrakaidecahedral cells on a BCC lattice the various elastic constants such as Young's modulus, shear modulus and Poisson's ratio can be calculated as a function of the density of the foam and edge cross-section at various conditions such as twisting, bending and the cell edge extension [46].

A micromechanical model of 3D open-cell foams with matrix method for calculating the elastic properties of the foam at three different loading rates [47].

Tarakanov et al. discussed a generalised equation for determining the stress-strain curves based on a cell having a 14-faced model. Mechanical properties calculated for the model fit with the experimental data. In the compression direction of [001] of Kelvin foam, they determined Young's modulus E_{100} for foams having low relative foam density which is given by equation [48].

$$E_{100} = \sqrt{2} E_p \frac{(b/L)^4}{2 + b/L} \tag{2.18}$$

Here, E_{100} is the polymer Young's modulus, b is the breadth of the square cross-section edge, and "L" is the edge length. The denominator in the above equation is a term for the correction of the volume of the vertex [48].

2.2.4 IRREGULAR MODELS

Vander Burg model: In this model, open-cell foams are explained using the geometry of a Voronoi tessellation [49]. A Voronoi tessellation has a list of points in space called sites. A Voronoi cell of each site is the region containing points closer to that site to others. A Voronoi tessellation of a symmetric collection of sites will share the symmetry.

Figure 2.9 shows a two-dimensional example. The left one is the Voronoi tessellation of a hexagonal lattice. The Voronoi cells are having the shape of regular hexagonal honeycombs. When the sites become slightly perturbed the tessellation will resemble the right one.

Various models are available but none of these models can be used for the exact modelling of foams.

Zhu et al. presented a method to produce a real open-cell foam. For an open-cell foam, the effective elastic properties can be determined with the implementation of micromechanical modelling based on Hill's lemma [50].

Bird et al. discussed the selection and design of foam for applications having energy absorption requirements with minimum cost and weight. The help of material indices developed based on energy absorption of various materials in this paper, opens a way for future works to quantify and compare the manufacturing costs of solid foam material having various relative densities [24].

Emanoil Linul et al. investigated the effect of various geometrical factors on the fracture toughness of open-cell foams using finite element modelling [51].

The work of Z. NIE et al. developed a 3D random Laguerre–Voronoi computational model for forms having open-cell structures. This model is quite close to the real foam topography. Hence it has more accurate results compared to the other open-cell foams models [52].

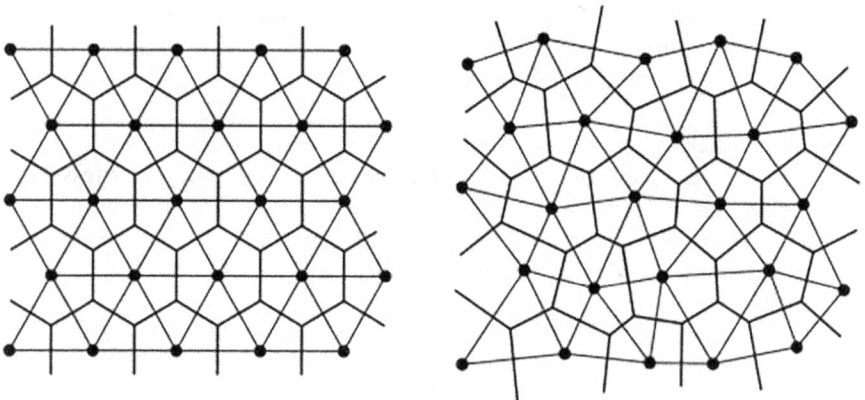

FIGURE 2.9 Two-dimensional Voronoi tessellations.

In the work of Ye et al., the foam cells have thin walls and uniform thickness these cells with polyhedric surfaces are observed in SEM and assumed the cell thin sheets having a similar thickness [36,53].

2.2.5   CLOSED CELL FOAMS AND THEIR APPLICABILITY

The gaps between the polymer materials cells in closed-cellular morphology are isolated by the cell walls. When compared with open-cell foams, the closed-cell foams have isolated cellular morphology and this provides better thermal insulation properties along with being lightweight and makes the material suitable for both aerospace and automotive industries. Thermal insulation, structural application, buoyance, etc. are the major application areas for closed-cell or rigid polymer foams [6].

Koohbor et al. discussed the dependence of the deformation given locally, on the instability caused by the cell wall, and the failure response of this on closed-cell foam. The researchers concluded that failure modes can be switched between elastic buckling and brittle fracture depending on the thickness of the foam cell wall and applied strain rate. At higher strain rates, the failure modes switched from elastic buckling and brittle fracture [54].

Andena et al. discussed the compression behaviour of expanded PP and polystyrene having different density ranges on the strain rate dependency. The microstructure and mechanical properties of these materials are examined by comparing the obtained experimental data with the existing foam models. The energy absorption diagram of both families of foams was obtained [55].

Andersons et al. explained the dependence of stiffness and strength on the anisotropy of rigid low-density closed-cell polyisocyanurate foams [56].

2.2.6   REGULAR MODELS

Gibson and Ashby model: The concept of the Gibson and Ashby model was already discussed in Section 3.1. But in the closed-cell foam, there is a small variation in the model compared to open-cell foam [20,38].

A typical schematic diagram of closed-cell foam with cubic unit cell arrangement having strut length, 'l', the thickness of struts t, and membrane thickness t_f is shown in Figure 2.10.

Goods et al. discussed closed-cell PU CRETE foam. They showed that for this closed-cell foam, the polymer material is distributed not only in the struts but also in the cell walls or faces also. If the thickness of the strut is t, the fraction of polymer in the struts is f, then the fraction of polymer in the cell wall with thickness t_f is 10 f [7,19].

The modelling of closed-cell foams is similar to open-cell foams. Here instead of struts having square cross-sections, we are using plates with square shapes that have side length of l and thickness as t [39].

$$\frac{\rho}{\rho_s} \propto \frac{t}{l} \qquad (2.19)$$

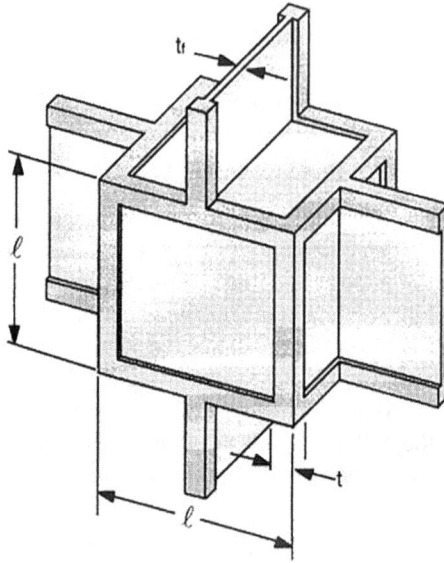

FIGURE 2.10 Closed cell unit cell having cubic symmetry [7,10,19,39].

$$I = \frac{lt^3}{12} \tag{2.20}$$

For a closed-cell foam, the stiffness is the sum of three components: the first component is the result of bending of the strut or bending of the edge and is also applicable to open-cell foams. The second term is due to the stretching of membrane or faces it is the result of strut flexure that causes the deformation of cell walls. The last term is the one related to internal gas pressure in closed cells [19].

The modulus of a closed-cell foam according to Gibson and Ashby is the sum of three components:

$$\frac{E^*}{E_s} \approx \varphi^2 \left(\frac{\rho^*}{\rho_s} \right)^2 + (1-\varphi)\frac{\rho^*}{\rho_s} + \left(\frac{P_0}{E_s} \right) \frac{(1-2\vartheta)}{1 - \left(\dfrac{\rho^*}{\rho_s} \right)} \tag{2.5}$$

The first component is the result of bending of the strut or bending of the edge and is also applicable to open-cell foams. The second term is due to the stretching of membranes or faces it is the result of strut flexure that causes the deformation of cell walls. The last term is the one related to internal gas pressure in the closed cells.

The contribution of struts of the cell to the modulus is given by the first term, the contribution of the cell wall of the foam is given by the second term and the final component is the result of internal gas pressure, where v* is Poisson's ratio of the foam [57].

2.2.7   IRREGULAR MODELS

Roberts A. P. et al. discussed a random model for determining the elastic properties of partially open cell foams and closed cell foams and partially open cell foams. They used finite element models for determining the experimental data and computed the Poisson's ratio, Young's modulus (E) etc. for the various random isotropic models. They compared the experimental data and computed data and these data are found to be similar [58].

Marvi et al. explained a modelling strategy for the determination of the mechanical behaviour of closed-cell rigid PU foams based on micromechanical characterisation and computational homogenisation [59].

Govaert et al. proposed a hybrid numerical experimental approach to determine the macroscopic mechanical behaviour of structured polymer foams. In this method, the microstructure of the foams is analysed with the three-dimensional X-ray computed tomography and is compared with an ideal representative volume element [60].

The paper of Barbier et al. derived a new law for determining the micromechanical behaviour under the uniaxial compression and tension of closed-cell Voronoi foams [61].

Marsavina et al. uses compression tests and digital image correlation to calculate the mechanical property experimentally for a PU foam of rigid structure. They investigate the effect direction of loading, density and rate of loading on the mechanical properties. To find out the elastic constants of the PU foam finite elemental analysis is done on a representative three-dimensional model of the foam [62].

2.3   CONCLUSION

Micromechanical modelling of both open-cell and closed-cell foams is discussed in this chapter. Over the past 50 years, there are various types of micromechanical models are developed for the modelling of polymeric foams. These models can either be the representation of a real foam structure (irregular) or an approximation (regular). But none of them is suitable to exactly describe the real foam properties.

These models can be used to predict the stress-strain responses of foam as a function of foam geometrical parameters, properties of the type of solid used, relative foam density and in the case of closed-cell foam gas pressure also. The strain energy in the foam when it is subjected to various strain rates can be used to fit the stress-strain predictions of the model. Currently, the foam models based on FEA have a great future to exactly determine the foam properties.

REFERENCES

1. M. Altan, "Thermoplastic Foams: Processing, Manufacturing, and Characterization," *Recent Res. Polym.*, 2018, doi: 10.5772/intechopen.71083.
2. T. -E. -E. Rostami, "Chemistry, Processing, Properties, and Applications of Rubber Foams," *Polymers,* vol. 13, pp. 1–53, 2021.
3. V. Shulmeister, Modelling of the Mechanical Properties of Low-Density Foams, *doctoral thesis,* Shaker Publishing BV, 1998, ISBN: 90-423-0025-6.

4. F. L. Jin, M. Zhao, M. Park, and S. J. Park, "Recent Trends of Foaming in Polymer Processing: A Review," *Polymers (Basel)*, vol. 11, no. 6, 2019, doi: 10.3390/polym11060953.

5. F. A. Shutov, "Foamed Polymers Based on Reactive Oligomers," *Adv. Polym. Sci.*, vol. 31, no. 39, pp. 1–64, 1981, doi: 10.1007/3-540-10218-3_1.

6. E. Burgaz, "Polyurethane Insulation Foams for Energy and Sustainability", *Adv. Struc. Mat.*, vol. 111, 2019, doi: 10.1007/978-3-030-19558-8.

7. V. Srivastava and R. Srivastava, "On the Polymeric Foams: Modeling and Properties," *J. Mater. Sci.*, vol. 49, no. 7, pp. 2681–2692, 2014, doi: 10.1007/s10853-013-7974-5.

8. Y. Chen, R. Das, and M. Battley, "Effects of Cell Size and Cell Wall Thickness Variations on the Strength of Closed-Cell Foams," *Int. J. Eng. Sci.*, vol. 120, pp. 220–240, 2017, doi: 10.1016/j.ijengsci.2017.08.006.

9. L. J. Gibson and M. F. Ashby , "The Mechanics of Three-Dimensional Cellular Materials",*Proceedings of the Royal Society of London. Series A, Mathematical and Published by : Royal Society Stable*. URL : https://www.jstor.org/stable/2, vol. 382, no. 1782, pp. 43–59, 1982.

10. M. Reiter and M. Jerabek, *Overview and Comparison of Modelling Methods for Foams*, 2020, doi: 10.1177/0021955X20966329.

11. J. Němeček, V. Králík, and J. Vondřejc, "A Two-Scale Micromechanical Model for Aluminium Foam Based on Results from Nanoindentation," *Comput. Struct.*, vol. 128, pp. 136–145, 2013, doi: 10.1016/j.compstruc.2013.07.007.

12. L. Marşavina and E. Linul, "Fracture Toughness of Rigid Polymeric Foams: A Review," *Fatigue Fract. Eng. Mater. Struct.*, vol. 43, no. 11, pp. 2483–2514, 2020, doi: 10.1111/ffe.13327.

13. L. J. Gibson and M. F. Ashby. *Cellular Solids, Structure and Properties*. Pergamon Press, New York, 1988.

14. R. Bashirzadeh and A. Gharehbagh, "An Investigation on Reactivity, Mechanical and Fire Properties of Pu Flexible Foam," *J. Cell. Plastics,* vol. 46, 2010, doi: 10.1177/0021955X09350805.

15. G. Harikrishnan, T. U. Patro and D. V. Khakhar, "Polyurethane Foam-Clay Nanocomposites: Nanoclays as Cell Openers," *Ind. Eng. Chem. Res.*, vol. 45, no. 21, pp. 7126–7134, 2006.

16. J. L. Throne, R. C. Progelhof, "Closed-Cell Foam Behavior Under Dynamic Loading-I. Stress-Strain Behavior of Low-Density Foams," *J. Cell. Plast.*, pp. 437–442, 1984.

17. L. F. Nielsen, "On Strength of Porous Material : Simple Systems and Densified Systems," *Mater. Struct.,* vol. 31, no. December, pp. 651–661, 1998.

18. J. A. F. Plateau, "Experimental and Theoretical Statics of Liquids Subjected to Molecular Forces Alone," *Mol. Forces*, vol. 2, 1873.

19. S. H. Goods, C. L. Neuschwanger, C. C. Henderson, and D. M. Skala, "Mechanical Properties of CRETE, A Polyurethane Foam", *J. Appl. Polym. Sci.,* vol. 68, pp. 1045–1055, 1998.

20. L. J. Gibson and M. F. Ashby, "Mechanics of Three-Dimensional Cellular Materials," *Proc. R. Soc. London, Ser. A Math. Phys. Sci.*, vol. 382, no. 1782, pp. 43–59, 1982, doi: 10.1098/rspa.1982.0088.

21. Y. Sun and Q. M. Li, "Effect of Entrapped Gas on the Dynamic Compressive Behaviour of Cellular Solids," *Int. J. Solids Struct.*, vol. 63, pp. 50–67, 2015, doi: 10.1016/j.ijsolstr.2015.02.034.

22. O. Lopez-pamies, P. Ponte, and M. I. Idiart, "International Journal of Solids and Structures Effects of Internal Pore Pressure on Closed-Cell Elastomeric Foams," *Int. J. Solids Struct.*, vol. 49, no. 19–20, pp. 2793–2798, 2012, doi: 10.1016/j.ijsolstr.2012.02.024.

23. P. Marcondes, Minimum Sample Size Needed to Construct Cushion Curves Based on The Stress-Energy Method, *MSC Thesis, Clemson University,* May 2007.

24. E. T. Bird, A. E. Bowden, M. K. Seeley, and D. T. Fullwood, "Materials Selection of Flexible Open-Cell Foams in Energy Absorption Applications," *Mater. Des.,* vol. 137, pp. 414–421, 2018, doi: 10.1016/j.matdes.2017.10.054.

25. K. Y. Jeong, S. S. Cheon, and M. B. Munshi, "A Constitutive Model for Polyurethane Foam with Strain Rate Sensitivity," *J. Mech. Sci. Technol.,* vol. 26, no. 7, pp. 2033–2038, 2012, doi: 10.1007/s12206-012-0509-1.

26. D. T. Morton, *Characterization and Modeling of the Mechanical Behavior of Polymer Foam,* 2021.

27. M. Avalle, G. Belingardi, and A. Ibba, "Mechanical Models of Cellular Solids : Parameters identification From Experimental Tests," *Int. J. Impact Eng.,* vol. 34, pp. 3–27, 2007, doi: 10.1016/j.ijimpeng.2006.06.012.

28. P. Prabhakar, H. Feng, and M. Doddamani, "Densification Mechanics of Syntactic Foam," *Compos. Part B Eng.,* vol. 232, pp. 1–14, 2020.

29. M. Avalle, G. Belingardi, and R. Montanini, *Characterization of Polymeric Structural Foams Under Compressive Impact Loading by Means of Energy-Absorption Diagram,* vol. 25, 2001.

30. F. Motor, "Load-Compression Behavior of Flexible Foams," *J. Appl. Polymer Sci.,* vol. 13, pp. 2297–2311, 1969.

31. A. M. Kraynik, M. K. Neilsen, D. A. Reinelt, and W. E. Warren, "Foam Micromechanics," *Foam. Emuls.,* pp. 259–286, 1999, doi: 10.1007/978-94-015-9157-7_15.

32. N. J. Mills, *Polymer Foams Handbook : Engineering and Biomechanics Applications and Design Guide,* no. April. 2007.

33. Q. Yu, B. E. Thompson and A. G. Straatman, *A Unit Cube-Based Model for Heat Transfer and Fluid Flow In,* vol. 128, no. April 2006, 2016, doi: 10.1115/1.2165203.

34. H. Haugen, J. Will, A. Köhler, U. Hopfner, j. Aigner and E. Wintermantel, "Ceramic TiO$_2$-Foams: Characterisation of a Potential Scaffold". *J. Eur. Ceram. Soc.,* vol. 24, no. 4, pp. 661–668, 2020.

35. Y. Maalej, M. I. El Ghezal, and I. Doghri, "Micromechanical Approach for the Behaviour of Open Cell Foams," *Eur. J. Eng. Mech.,* vol. 22, pp. 198–208, 2013.

36. R. B. Pecherski, "Macroscopic Properties of Open-Cell Foams Based on Micromechanical Modelling," *Eur. J. Eng. Mech.,* vol. 23, pp. 234–244, 2003.

37. T. Daxner, R. D. Bitsche, and H. J. Böhm, "Space-Filling Polyhedra as Mechanical Models for Solidified Dry Foams," *Mater. Trans.,* vol. 47, no. 9, pp. 2213–2218, 2006, doi: 10.2320/matertrans.47.2213.

38. P. L. J. Gibson, M. F. Ashby, G. S. Schajer and C. I. Robertson, "The Mechanics of Two-Dimensional Cellular Materials", *Proceedings of the Royal Society of London. Series A, Mathematical and Published by : Royal Society Stable,* vol. 382, no. 1782, pp. 25–42, 1982.

39. R. E. M. M. M.F. Ashby, "The Mechanical Properties of Cellular Solids," *Metall. Trans. A,* vol. 14, no. September, pp. 1755–1769, 1983.

40. S. U. Timoshenko, *Elements of Strength of materials 4th Edition.* East-West Press Private Ltd, New Delhi, 1960.

41. W. Thomson, "On the Division of Space with Minimum Partitional Area," *Acta Mathematica,* vol. 11, pp 121–134, 1888.

42. J. Storm, M. Abendroth, and M. Kuna, "Influence of Curved Struts, Anisotropic Pores and Strut Cavities on the Effective Elastic Properties of Open-Cell Foams," *Int. J. Mech. Mater.,* 2015, doi: 10.1016/j.mechmat.2015.02.012.

43. J. Storm, M. Abendroth, and M. Kuna, "Numerical and analytical solutions for aniso-tropic yield surfaces of the open-cell K ELVIN foam," *Int. J. Mech. Sci.,* vol. 105, pp. 70–82, 2016, doi: 10.1016/j.ijmecsci.2015.10.014.

44. C. Ge and D. Cormier, "A Preliminary Study of Cushion Properties of a 3D Printed Thermoplastic Polyurethane Kelvin Foam," *Pack. Technol.*, vol. 2016, pp. 1–8, 2017, doi: 10.1002/pts.2330.

45. L. Kucherov and M. Ryvkin, "Fracture toughness of open-cell Kelvin foam," *Int. J. Solids Struct.*, vol. 51, no. 2, pp. 440–448, 2014, doi: 10.1016/j.ijsolstr.2013.10.015.

46. H. X. Zhu, J. F. Knott, and N. J. Mills, "Analysis of the elastic properties of open-cell foams with tetrakaidecahedral," *J. Mech. Phys. Solids*, vol. 45, no. 3, pp. 27, 1997.

47. K. Li, X. Gao, and A. K. Roy, "Micromechanical Modeling of Three-Dimensional Open-Cell Foams Using the Matrix Method for Spatial Frames," *Compos. Part B: Eng.*, vol. 36, pp. 249–262, 2005, doi: 10.1016/j.compositesb.2004.09.002.

48. A. G. D. and O. G. Tarakanov, "Effect of Cellular Structure on the Mechanical Properties of Plastic Foams," *Polymer Mech.*, vol. 4, pp. 519–525, 1973.

49. R. M. M. W. D. Van Der Burg, V. Shulmeister, E. Van Der Geissen, "On the linear elastic properties of regular and random open cell foam models," *Theory Psychol.*, vol. 12, no. 6, pp. 825–853, 2015.

50. W. Zhu, N. Blal, S. Cunsolo, and D. Baillis, "AC US CR," *Int. J. Solids Struct.*, 2017, doi: 10.1016/j.ijsolstr.2017.02.031.

51. E. Linul and L. Marsavina, "Influence of Geometrical Parameters on Fracture Toughness for Open Cell Foams", *19th International Conference Computer Methods in Mechanics*, pp. 3–4, 2011.

52. Q. T. Nie, Z. Y. Lin, "The Work Reports on the Development of Random Three-Dimensional Laguerre-Voronoi Computational Models for Open Cell Foams. The Proposed Method Can Accurately Generate Foam Models Having Randomly Distributed Parameter Values. A Three-Dimensional Model of c," *Arch. Metall. Mater.* vol. 63, pp. 1153–1165, 2018, doi: 10.24425/123788.

53. W. Ye, C. Barbier, W. Zhu, A. Combescure, and D. Baillis, "Macroscopic Multiaxial Yield And Failure Surfaces For Light Closed-Cell Foams," *Int. J. Solids Struct.*, vol. 69–70, pp. 60–70, 2015, doi: 10.1016/j.ijsolstr.2015.06.008.

54. B. Koohbor, S. Ravindran, and A. Kidane, "Effects of Cell-Wall Instability and Local Failure on The Response of Closed-Cell Polymeric Foams Subjected to Dynamic Loading," *Mech. Mater.*, vol. 116, pp. 67–76, 2018, doi: 10.1016/j.mechmat.2017.03.017.

55. L. Andena, F. Caimmi, L. Leonardi, M. Nacucchi, and F. De Pascalis, "Compression of Polystyrene and Polypropylene Foams for Energy Absorption Applications : A Combined Mechanical and Microstructural Study, " *J. Cell. Plast.*, 2019, doi: 10.1177/0021955X18806794.

56. J. Andersons, M. Kirpluks, L. Stiebra, and U. Cabulis, "Anisotropy of the Stiffness and Strength of RIGID Low-Density Closed-Cell Polyisocyanurate Foams," *Mater. Des.*, vol. 92, pp. 836–845, 2016, doi: 10.1016/j.matdes.2015.12.122.

57. R. S. Lakes, "Cellular Solids," *J. Biomech.*, vol. 22, no. 4. p. 397, 1989, doi: 10.1016/0021-9290(89)90056-0.

58. A. P. Roberts and E. J. Garboczi, "Elastic Moduli of Model Random Three-Dimensional Closed-Cell Cellular Solids," *Acta Mater.*, vol. 49, no. 2, pp. 189–197, 2001, doi: 10.1016/S1359-6454(00)00314-1.

59. M. Marvi-mashhadi, C. S. Lopes, and J. Llorca, "Modelling of the Mechanical Behavior of Polyurethane Foams by Means of Micromechanical Characterization and Computational Homogenization," *Int. J. Solids Struct.*, vol. 146, pp. 154–166, 2018, doi: 10.1016/j.ijsolstr.2018.03.026.

60. J.G.F. Wismans, J.A.W. Van Dommelen, L.E. Govaert, H.E.H. Meijer and B. Van Rietberg, "Computed Tomography-based Modeling of Structured Polymers," *J. Cell. Plast.*, vol. 45, 2009, doi: 10.1177/0021955X08100045

61. C. Barbier, P. M. Michaud, D. Baillis, J. Randrianalisoa, and A. Combescure, "New Laws for the Tension/Compression Properties of Voronoi Closed-Cell Polymer Foams in Relation to their Microstructure," *Eur. J. Mech./A Solids*, vol. 45, pp. 110–122, 2014, doi: 10.1016/j.euromechsol.2013.12.001.

62. L. Marsavina, D. M. Constantinescu, E. Linul, T. Voiconi, D. A. Apostol, and T. Sadowski, Foams Fracture toughness of rigid polymeric foams: A review," *Fatig. Fract. Eng. Mat. Struct.",* vol. 43, pp. 1–32, 2020, doi: 10.1111/ffe.13327.

3 Latex-Based Polymeric Foams – Preparation, Properties and Applications

Leny Mathew

CONTENTS

3.1 INTRODUCTION

Latex foam rubber is defined as a cellular rubber made directly from liquid latex in which all the cells are intercommunicating and have a vulcanised cellular structure. The smooth surface of the foam rubber which is formed by contact with the surface of the mould is called the skin and the foam rubber is mainly used as cushioning material and in household applications. It is also used in automobiles and railways as

DOI: 10.1201/9781003218692-3

a cushioning material. The advantages of latex foam rubber compared to other foam materials are better. The foam rubber is soft to touch, resilient and got damping and fatigue resistance. The easiness of moulding into a variety of shapes, sizes and hardness makes it more industrialised and economic.

The development of 'Dunlop Process' for the manufacture of foam rubber in 1920 opened the present era of the foam industry. In this process, foamed latex compound is allowed to set in mould using a delayed action gelling agent – sodium silicofluoride. Now the process is mainly used for the production of foam mattresses, cushions, pillows and many foam rubber products.

In the late 1950s, the Talalay Process for the manufacture of foam rubber was invented. In Talalay process, the mechanically partially foamed latex compounded is expanding by the application of vacuum and the expanded foam is subjected to rapid freezing technique. Then carbon dioxide is permeated through the frozen foam. The foam is then vulcanised and the product is stripped off from the mould and dried. The unique freezing technique of Talalay process prevents particles from settling and ensures consistent cell structure to the foam rubber.

There are two interfaces in the foamed latex: serum–rubber interface and serum–air interface. The successful manufacture of latex foam depends upon the manipulation of these two interfaces.

The destabilisation of the serum–air interface is preceded by that of the serum–rubber interface.

The serum–rubber interface is destabilised so that the rubber comes together and forms a reticulate structure. The foam should be solidified before starting the tendency to collapse. This is achieved through the judicious choice of foam promoter, gelling agent and foam stabiliser [1].

3.2 MANUFACTURE OF LATEX FOAM BY DUNLOP PROCESS

The process comprises the following major steps [1–3]

1. Compounding the latex with various compounding ingredients to meet the service requirements.
2. The compounded latex is foamed with the assistance of added foaming agents.
3. Transfer the foamed latex into a mould.
4. The latex foam in the mould is allowed to gel at normal ambient temperature.
5. The gelled foam is then vulcanised followed by washing and drying.

3.2.1 PREPARATION OF LATEX COMPOUND

3.2.1.1 Raw Latex

Natural Rubber Latex and SBR latex can be used for latex foam production. Centrifuged Natural Rubber latex is mainly used for the production of latex foam. The latex is deammoniated to 0.2–0.25% ammonia content before compounding and processing. For the production of foam rubber, the quality of the raw latex should be good. Certain parameters such as odour, magnesium content, VFA and KOH no. Have much effect on quality and production process. Natural rubber latex foam rubber is not flame resistant. The flame resistance of natural rubber latex foam rubber can be improved by suitably compounding natural latex with flame retardants like antimony trioxide chlorinated paraffin wax, zinc borate or hydrated aluminium oxide.

3.2.1.2   Foam Promoters

Foam promoters are added to natural rubber latex in order to promote rapid foaming. The mixture of carboxylate soaps promotes foaming more effectively. For natural rubber latex, the amount of foaming agents varies from 0.2 to 2 phr. Potassium oleate soap functions both as a stabiliser and as foaming agents.

3.2.1.3   Foam Stabilisers

Foam stabilisers are also known as Secondary gelling agents. Several foam stabilisers are being used in foam production. Foam stabilisers comprise mainly cationic soap, polyamines and guanidines. By the reaction of cationic soap with the anionic protective layer on the rubber particles, a reduction in mechanical stability is affected. This will lead to improved gelling, a rise in gelling pH and consequently improved foam structure. Among the quaternary ammonium soaps, cetyle trimethyl ammonium bromide is widely used. Poly ammines such as triethylene tetramine and tetra ethylene pentamine are used as foam stabilisers. Diphenyl guanidine (DPG) is also widely used as a foam stabilising agent. Generally, the dosage of the foam stabiliser is one-half by weight of the overall amount of carboxylate foaming agents used.

3.2.1.4   Delayed Action Gelling Agents

The main characteristic of the delayed action gelling agent is that when it is added to a latex its action is time dependent. The destabilisation of the latex compound takes place after a lapse of a certain time and gellation takes place. The most important delayed action gelling agent is the alkali metal salt of hydro fluoro silicic acid and such salts are commonly known as silicofluorides.

The hydrolysis of sodium silicofluoride is

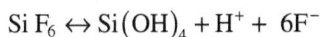

$$Na_2SiF_6 \leftrightarrow 2Na + Si\,F_6$$

$$Si\,F_6 \leftrightarrow Si(OH)_4 + H^+ + 6F^-$$

The gellation with sodium silicofluoride takes place at an alkaline pH of around 8.5.

Apart from the destabilisation action of hydrofluoric acid, it also plays a role in gelation.

When zinc oxide is added to the latex compound the ammonium fluoride resulting from the hydrolysis of the sodium silicofluoride in the presence of ammonia reacts with zinc oxide and ammonia to form zinc ammines which results in the destabilisation of the latex compound.

There are four main actions involved in the gelation process.

1. Gradual reduction of pH by the liberation of hydrogen ions.
2. The latex stabilisers become absorbed into the precipitated silica.
3. Co-precipitation of latex particles in the gelatinous silica which forms *insitu*.
4. Due to the lowering of the pH the concentration of dissolved zinc increases and this leads to the formation of insoluble zinc soaps at the surface of the rubber particles.

Sodium silicofluoride is used in conjunction with zinc oxide and for obtaining a firm coherent gel, the addition of Zinc oxide is essential. Sodium silicofluoride is prepared as a 50% dispersion in water using Dispersol F and it is added to the latex compound at 20% concentration. It is desirable to adjust the pH of the diluted dispersion to about 6 before the addition.

3.2.1.5　Cure Systems

A conventional cure system [3–5] is normally used for the manufacture of foam rubber from natural rubber latex and the sulphur is used at the level of 2–2.5 phr. The accelerators used are zinc diethyldithiocarbonate together with a small quantity of zinc-2 mercaptobenzothiazole and this combination imparts better compression modulus and lower compression set.

3.2.1.6　Fillers

In latex foam production, fillers are used in order to stiffen the rubber phase of the foamed rubber. The other advantage obtained from the use of fillers is to reduce shrinkage, which occurs during the washing and drying of the foam products. Commonly china clay or whiting is used and the dosage is up to 30 phr.

3.2.1.7　Flame Retardants

Flame resistance of the natural rubber latex foam can be improved by adding chlorinated paraffins, antimony tri oxide, zinc borate or hydrated aluminium oxide.

Deammoniated latex is compounded by adding corresponding compounding ingredients and compounding is done in two stages (Table 3.1).

In batch-wise foam production, simple machinery such as a Hobart mixer is used. The mixer comprises a bowl in which a wire whip is rotated in planetary motion. Air is introduced into the compounded latex by the whipping process.

The process flow chart is given below

After the addition of first-stage compounding ingredients into deammoniated latex, the compound is kept for maturation for a period of 24 h.

3.2.2　Foaming of Compounded Latex

Foaming of the above latex compound is achieved by mechanical agitation in the presence of air in the planetary mixer. For this potassium oleate is first added into the latex compound, as per 2nd stage compounding and it is added into the bowl of

TABLE 3.1

Basic formulation of compounding

Sl. No	Ingredients	Parts by Weight	
		Dry	**Wet**
	Ist stage compounding		
1	Natural Rubber Latex (60%)	100	167
2	Potassium oleate (20%)	0.2	1.0
3	Sulphur (50%)	2.5	5
4	ZDC (50%)	1.0	2.0
5	ZMBT (50%)	0.5	1.0
	IInd stage compounding		
6	20% Potassium oleate solution	1.0	5.0
7	Fillers (as paste)	As desired	
8.	DPG (50%)	0.6	1.2
9	Zinc Oxide (50%)	4	8
10	20% sodium silicofluoride dispersion	1.5	7.5

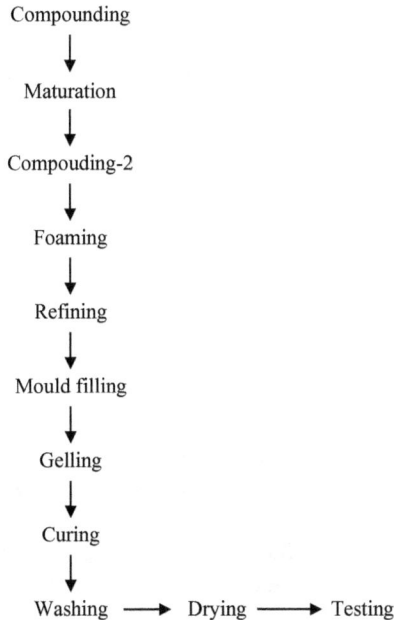

Compounding
↓
Maturation
↓
Compouding-2
↓
Foaming
↓
Refining
↓
Mould filling
↓
Gelling
↓
Curing
↓
Washing ⟶ Drying ⟶ Testing

a planetary mixer. The whip is rotated at a high speed to expand the foam and raise it to the required height as rapidly as possible. The density of the finished product depends directly on the fourth density and there the fourth height. After attainment of the required height, the whipping speed is reduced to comminute the bubbles and refine forth without further reducing the density. Just before the required volume is reached, filler paste is added.

3.2.3 ADDITION OF GELLING AGENTS

Zinc oxide and sodium silicofluoride are used as gelling agents, and DPG is used as secondary gelling agent. A measured amount of zinc oxide dispersion and DPG dispersion is added to the bowl and stirred into the fourth followed by a measured quantity of sodium silicofluoride dispersion. After a further period of blending the fourth is poured from the bowl into a mould.

3.2.4 MOULDING OF THE FOAM

The mould used is usually made from cast aluminium, the lids containing the domed cavity formers. The cast aluminium alloy moulds combine the desired properties of good heat transfer resistance to corrosion, lightweight and smooth surface to facilitate stripping. Each mould is conditioned to ensure that its temperature is uniform and constant generally about 40–50°C and treated with a mould-releasing agent. The most widely used mould release agents are silicone emulsion, aqueous solutions of sodium carboxy methyl cellulose or medium molecular weight polyethylene glycols or a mixture of both.

To ensure good moulding, the mould release solution should be applied evenly and uniformly over all surfaces and allowed to dry. To get the best result, the mould release agent is sprayed immediately after demoulding a product when the mould is still hot after curing. The mould release solution dries off quickly and serves also as an aid to mould cooling.

3.2.5 FOAM GELLING

Sensitised foamed latex is poured into the treated mould and positioned on the lid. A short interval is allowed before the mould passes into the curing chamber. During this time, the zinc oxide and sodium silicofluoride addition bring about the gellation of the foam. The addition of gelling agent is adjusted so that gelation is complete in about 5–10 min. Control of gelling reactions is important for good-quality foam production. The level of gelling agents and their ratio is important in balancing and controlling the reactions. Mould temperatures are also important and these are usually controlled to 30–35°C for the pan and 40–45°C for the lids. Failure to control foam temperature and gelling agent additions can lead to poor foam structure throughout the products. Failure to maintain mould temperature can lead to defects such as thick skin or loose skin on the surface of the product.

3.2.6 FOAM CURING

Following the gelation, the mould passes into a curing chamber where they are exposed to live steam at atmospheric pressure for 20–30 min. The time depends on the thickness of the product. Thin sheets require less time and thick sheets require more time. After that, the curing lid is opened and the products are removed from the mould.

3.2.7 Washing and Drying

The foam product as it is removed from the mould contains a number of water-soluble non-rubber constituents, such as soap, which could lead to poor ageing, poor resilience and bad odour. These contaminates are simply and effectively removed by passing the moulding through a series of rollers where clean water is sprayed onto the foam. The excess of water at the end of the washing stage is removed by passing through a final squeeze roller. The final product is then kept for drying. Drying is done at 70°C for 3–4 h.

The dries for latex foam products are continuous with rapid air circulation. The drying rate is governed to a large extent by the speed with which water in the product can migrate to the surface, where it is evaporated.

The wet products are placed in an open mesh conveyor which passes through a camber with hot air circulation. The air is directed by fans onto the upper and lower surface of the products so that there is a positive pressure on the upper surface to prevent the products from being blown out of the air. Temperature is generally between 100 and 110°C and drying times are normally about 1–1 ½ h.

3.2.8 Defects and Remedies in Foam Rubber

The common defects encountered in the manufacture of foam rubber are

1. **Shrinkage**
   Normal shrinkage in natural latex foam rubber on drying is around 10%. High shrinkage may be due to a lack of maturation of latex compound or an improper gelling system. It is also due to high expansion in volume and low filler level.
2. **Foam Collapse**
   It is due to poor foam stability. Increasing foam stabilisation and gelling agent are the suitable solutions for this.
3. **Coarse structure and rat holes**
   It occurs due to gelation at low pH. Increasing the secondary gelling agent and adjusting the refining time are remedies for this.
4. **Loose skin**
   It is due to cold mould. Mould temperature is to be adjusted to 38–40°C
5. **Thick Skin**
   It is due to high mould temperature. Use cooler mould.
6. **Flow mark and splitting in centre**
   It is due to fast gellation. Adjust gelling system.

3.2.9 Testing and Grading of Latex Foam Rubber

Foam rubber is tested as per BIS 1741-1960. The important properties tested are described below [6–8].

TABLE 3.2

The foam rubber products are graded by the indentation hardness index

Grades	Indentation hardness index
Soft	7–20
Medium	21–35
Hard	36–50
Extra Hard	51 and above

1. Indentation Hardness Index

It is a partial measure of comfort and it is the load in kilograms required to give an indentation in the sample equivalent to 40% of the original thickness of the sample with a specified indent (300 mm diameter).The hardness of the product depends upon the density (Table 3.2).

2. Compression Set

This property is a measure of the state of cure of the foam. The test consists of maintaining the test piece under specified conditions of time, temperature and constant deflection and the effect on the height of the released test piece.

BIS standards specify two methods for the determination of compression set

Method A

Sample test piece of 5 ×5 ×2.5 cm dimension is cut from the foam sample. Measure accurately to 0.1 mm initial thickness of the test piece using a dial gauge. It is then placed in the compression device and compressed to 50% of its thickness between the parallel steel plates, using steel spacers between the plates. Within 15 min, place the compressed test piece in an air oven at 70°C for 22 h. Remove the test piece from the oven and release it from the compression at the earliest. Allow to recover for 30 min at room temperature and remeasure the thickness.

Compression set is calculated as follows

$$\text{Compression set } (\%) = \frac{\text{To} - \text{Tr}}{\text{To}} \times 100$$

Where To = Initial thickness of the test piece in mm.
Tr = Thickness of the test piece in mm after recovery.

Method B

Measure the original thickness of the whole sample. Place it under a circular flat indentor, 305mm in diameter and weighing 25 kg on the middle of the sample and allow it to remain thereon for 168 h at room temperature and humidity. After this period remove the load and allow the sample to recover for 30 min. Measure any permanent depression produced and calculate compression set as follows:

$$\text{Compression set } (\%) = \frac{\text{Change in thickness } \times 100}{\text{Initial thickness}}$$

3. Colour

The colour of the latex foam rubber products shall be agreed upon between the purchaser and the supplier.

4. Odour

The latex foam products shall have no objectionable odour

5. Flex Resistance

When a latex foam rubber product is subjected to the flexing test, its indentation hardness shall be not reduced by more than 20% and the thickness by more than 5% of the original hardness and thickness respectively. The flexing test involves submitting the whole sample to continuous flexing with an indentor for 2,50,000 cycles per second and measuring the loss in hardness and thickness.

6. Metallic impurities

The copper and manganese content in the sample shall not be more than 0.001% and 0.005%, respectively.

7. Ageing

The test sample is kept in an oven maintained at 70 + 1 º C for 168 h. Hardness of the unaged and aged shall not vary by more than 20%.

3.2.10 APPLICATION OF LATEX FOAM RUBBER

Foam rubber is found in a wide range of applications, from cushioning in automobile seats, bedding and furniture to insulation in walls and appliances to soles and heels in footwear. The advantages of latex foam rubber compared to other foam materials are

1. Softness to touch
2. Resilience
3. Damping and fatigue resistance.
4. Easiness for moulding into a variety of shapes, sizes and densities

3.3 CONCLUSION

Latex foam rubber is defined as a cellular rubber made directly from liquid latex in which all the cells are intercommunicating and have a vulcanised cellular structure. The smooth surface of the foam rubber which is formed by contact with the surface of the mould is called the skin and the foam rubber is mainly used as cushioning material and in household applications. It is also used in automobiles and in railways as a cushioning material. The advantages of latex foam rubber compared to other foam materials are better. The foam rubber is soft to touch, resilient and got damping and fatigue resistance. The easiness of moulding into a variety of shapes, sizes and hardness makes it more industrialised and economic.

REFERENCES

1. Blackey, D. C. (1997). *High polymer lattices, science and technology*, vol. 2, 2nd ed. London: Chapman & Hall.
2. The Malaysian Rubber Producers' Research Association. (1984). *The natural rubber formulary and property*. London: Imprint of Luton Ltd.
3. Chapman, A. V. & Porter, M. (1988). *Sulphur vulcanization chemistry*. Oxford: Oxford University Press.
4. Amir Hashim, M. Y. & Moris, M. D. (1999). NR latex vulcanization – Prevulcanization and postvulcanization of dipped NR latex films. *J. Rubb. Res.*, 2(2), 78–87.
5. Ma'zam, M.S., Dazylah, D. & Mok, K. L. (2006). Recent development of peroxide prevulcanized NR latex. *Paper presented at the 3rd International Rubber Glove Conference*, 12–14 September, Kuala Lumpur.
6. Ng., K. P., Ma'zam, M. S., Lai, P. F. & Abu, A. (2003). *Development of low protein lattices. RRIM Monograph no. 17*. Kuala Lumpur: Rubber Research Institute of Malaysia.
7. American Society for Testing and Materials (ASTM). (2009). Test method for colorimetric/spectrophotometric procedure to quantify extractable chemical dialkyldithiocarbamate, thiuram, and mercaptobenzothiazole accelerators in natural rubber latex and nitrile gloves. *Am. Soc. Testing Mat.*, 12, 1–12.
8. American Society for Testing and Materials (ASTM). (1995). Standard test method for analysis of protein in natural rubber and its products. *Am. Soc. Testing Mat.*, 110, 1–7.

4 Blended Polymeric Foams

Tharun A. Rauf and Athira C. J.

CONTENTS

4.1 INTRODUCTION

The present world is giving much importance to polymer alloys, blends, and composites. These materials were given prominence not to substitute them with wood or any other materials, but to enhance the properties of the traditional polymers. Betterment means providing better materials that are economically viable and have superior properties. Polymer blend technology has evolved in such a manner that, it is being used as commodity resins, engineering resins, and specialty resins. Most of the polymer pairs are immiscible in nature and hence the method of preparation won't be spontaneous in

DOI: 10.1201/9781003218692-4

nature. Hence the method of preparation is an important parameter for polymer blends. Many types of polymer blending processes are there – especially by physical methods that include different types of mixing process. The advantages of blending are improvement of resin properties and processability. The processability factor is an important parameter in the making of polymer blend foams. Inclusion of elastomeric particles convalesces gas bubble nucleation, hence stabilizing the process of foaming. Bubble size can be reduced and thus the final foam density. Polymer blends are capable of forming homogenous foam. The interface area between the different phases can be used as bubble nucleating agents during the course of the foaming process.

For the last ten years, the foamed polymeric material has progressed into a prominent technology for the synthesis of cellular compounds with the required properties. Consequently, nowadays, foams are overpowering many applications that include load bearing, energy absorption, thermal insulation, packaging, and so on. Conversely, the demand for refined, controlled cell morphologies, and enhanced mechanical properties is growing at a greater pace. Even bio-based polymer foams are also being developed since economically viable polymer foams are demanded the most. Previously porous materials were synthesized primarily using organic solvents, which were later proved to be against green synthetic phenomena. Hence, for the foaming process, suitable gas was selected as the blowing agent which is suitable for the making of foamed polymer.

The polymers that are generally used for the preparation of blend foams are co-polymers of poly phenylene ether (PPE) and Styrene Acrylo Nitrile (SAN), Polystyrene and SAN [1], Ethylene– propylene-diene (EPDE) and Polypropylene (PP) [2], poly butylene succinate (PBS) and Polylactic acid (PLA), PLA/PBS/Activate Carbon(AC) [3], Thermoplastic gelatin (TPG)/PBS, Polybutylene Terephthalate (PBT), High-Density Polyethylene (HDPE) and PBT, and Polytetrafluoro ethylene (PTFE) and PBT [4].

For the making of biodegradable type polymer blend foams, blending of polymers was made by the combination of PBS/Polybutylene adipate-co-terephthalate [4]. It was then foamed by blowing with CO_2. As a continuation of this criteria, blends of PP and Wood flour composites were also prepared.

4.2 BLOWING AGENTS

Porous nature can be obtained for the blends either by using the organic solvents or by using the blowing agents. The latter was found to be more economically and environmentally viable. Also, the production cost was found to be less. Two types of blowing agents are typically used, Azodicarbonamide (ADC) and CO_2.

ADC is an established blowing agent for foam production. Numerous scientists are utilizing ADC for PBS [5–9]. But some researchers have explained that heat release from the exothermic reaction of ADC during the foaming process can cause some unexpected changes due to the low viscosity of linear PBS. The foamed polymer may experience cell mishap, thus deteriorating mechanical or insulation properties. Hence, the required property and preferred application will not be nearby during this process.

Hence, it can be attributed that usage of CO_2 as a blowing agent will be the most suitable one since it won't cause any adverse effects on the compounds. But the blended polymer should be susceptible to CO_2 blow.

4.3 DIFFERENT PREPARATIONS OF POLYMER BLEND FOAMS

4.3.1 BASICS OF FOAMING

For understanding the structural/morphological modification of foams and their different processing techniques, basic knowledge about the different foaming processes involved has to be understood. A literature survey will do the help [10,11].

The process of foaming involves six steps:

1. Development of a homogenous polymer/Gas mixture
2. Cell nucleation
3. Cell emergence
4. Stabilization of cell
5. Melt-blended compounds
6. Beaded blends

4.3.1.1 Development of Homogenous Polymer/Gas Mixture

Using a blowing agent homogenization of the polymer is carried out, which has been determined by mass-transfer process. Here, the blowing agent is diffused into the melt or solid bead and contained in the polymer-gas solution. The temporal and spatial dependency of this transport is determined by Fick's law. Apart from temperature, diffusion of the blowing agent is also depending on pressure and gas concentration in the polymer [12]. Hence, it can be stated that the diffusion of gas into the bead depends on time and space. For diffusion, the free volume of the polymer is important [4] since the increase in free volume increases diffusivity. In juxtaposition to diffusion, the incorporation of gas into the polymer matrix depends on the applied pressure. As per Henry's law, concentration of a dissolved blowing agent is totally depending on the pressure applied, in polymeric melt. But in terms of temperature, it is negative exponential. With the increase in temperature, the solubility of gas is reduced. In addition to that higher shear rates will also reduce the solubility of the blowing agent in the polymer matrix. It is mainly due to the decrease in the free volume, which is mainly caused by the polymer chain alignment [13].

4.3.1.2 Cell Nucleation

Nucleation is the creation of nuclei, which act as the midpoint for cell growth after a drop in pressure. It can take place either in an underwater pelletizing system or in an autoclave. Simply the nucleation process is expressed in Figure 4.1 which will explain how the nucleation process is taking place.

The above figure depicts the nucleation process. When it comes to the nucleation process on a wide scale, even though the formation is the same, the methodology imparted will be different. A sudden pressure drop can cause a reduction of solubility, which in turn can develop a driving force to reduce the gas content of the polymer gas mixture. If the temperature has increased in a sudden manner, then also solubility can be achieved. Nucleation process can be manipulated either through homogeneous or heterogeneous modes. The difference lies in the condition that homogeneous nucleation occurs away from the surface of the system, whereas

FIGURE 4.1   Nucleation formation.

heterogeneous nucleation occurs at the surface of the system. It has been observed that heterogeneous nucleation process occurs much more rapidly than the homogeneous nucleation process. Heterogeneous nucleation applies to the phase transformation between any two phases of gas, liquid, or solid, typically, for example, condensation of gas/vapor, solidification from liquid, and bubble formation from liquid. According to nucleation theory, nucleation starts at clusters of gas molecules inside the melt [14]. The voids formed due to nucleation will be acting as nucleating sites. The homogeneous nucleation rate is heavily dependent on the pressure drop. A high-pressure drop precedents a high nucleation rate. But the rate of nucleation is depending on the surface tension between polymer-melt and gas.

Foaming of Blended PP with PBT/PTFE

Blending of PBT with PP can yield a better product since PBT is the heterogenous nucleating agent. The physical network formed by the entanglement of fibrous PBT can effectively restrict the growth of bubbles in the foaming process, helping to reduce cell size as well as improve cell density and uniformity [15]. In comparison with HDPE and PBT, the fibrous PTFE can enhance the CO_2 sorption capacity of the matrix. Also, it will form a physical network in the melt to restrain the bubbles. Hence, the product obtained has shown a great increase in bubble density and volume expansion ratio [16,17]. The structures were prepared by batch foaming process.

4.3.1.3   Cell Emergence

As per nucleation theory, nucleation starts at clusters of gas molecules inside the melt [17]. Those voids act as nucleating sites. The homogeneous nucleation rate is purely dependent on the decrease in pressure. A high-pressure decrease leads to an increase in nucleation rate. Surface tension between polymer melt and gas is the key factor that depends on the nucleation rate, which has been explained earlier.

The nucleation process is employed generally in thermoplastic materials. In this foam extrusion process, a physical foaming agent is involved, which is one of the key steps in designing and optimization of the process. The conditions that prompt phase separation and bubble nucleation are obviously connected to the solubility parameters, i.e., temperature and pressure at a given foaming agent content, conditions that can be probed during degassing experiments. Addition of a nucleating agent can substantially modify these degassing conditions.

4.3.1.4 Stabilization of Cells

Another type of nucleation process is stress, both in elongation and shear [15,16]. Extensional stress around growing cells is responsible for pressure fluctuations, which reduces solubility and thereby increases super-saturation [17]. Shear initiates micro-voids and causes an elongation of already existing bubbles. These types of mechanisms lead to an increased nucleation rate and greater cell densities. After the nucleation process, cell growth is taking place. During cell growth, the stored gas diffuses out of the melt to the nucleation sites. The impetus behind this process is super-saturation caused by the pressure drop or temperature increase. Hence, it can be stated that this process is temperature dependent, which in turn influences diffusion, pressure drop rate, and the actual pressure [17]. Another foaming parameter is the visco-elastic properties of the melt since the melt is subjected to elongational deformation during bubble growth [18]. To obtain foams with a favorable cell size, cell size distribution, and thereby good properties (e.g. mechanical behavior or thermal transport properties), the morphology must be stabilized and cell growth has to cease, otherwise cell coalescence or coarsening (large cells grow at expense of small ones) takes place and deteriorates the final foam morphology. The main factor for stabilization is reducing the polymer's temperature and thus the melt's viscosity increases. As the blowing agent is diffusing out of the polymer, the viscosity increases even further, since dissolved gas in a polymer may act as plasticizing agent [19]. At large elongations, strain-hardening is important. It raises elongational viscosity above the linear value due to the stretching of chains [20]. Due to strain-hardening, thin sections of a cell wall are more difficult to extend than thicker ones (thinner sections are subjected to higher strains and thereby a higher degree of stretching of chains). So the thick sections are extended preferentially, known as the self-healing effect [21]. Strain-hardening can be induced by long chain branching [22], the introduction of high-aspect-ratio nano-additives [23], or by blends having fibril morphology [24]. Effects of countering cell stabilization are the creation of crevices by large and fast deformations and rupture of cell walls [10].

4.3.1.5 Melt Blended Polymer Foams

It is the most preferred method for the preparation of clay/polymer nanocomposites of a thermoplastics and elastomeric polymeric matrix. Usually, the polymer is melted and combined with the desired amount of the intercalated clay using Banbury or an extruder. A twin screw extruder is normally employed for this process. Compression molding machines can also be utilized for the same, provided the application of the compound has to be analyzed.

Manufacture of PPE-co-SAN Foams

Initially, PPE and SAN [2] were dried in a vacuum oven for 24 h at 80°C. Janus particles (JP), was used as compatibilizer. To prevent further cross-linking, the uncross-linked double bonds in the PB core of the JPs, the latter were dried at 40°C in a vacuum. The blend systems were compounded with a co-rotating twin-screw extruder with a screw diameter of 20 mm and a screw length of 600 mm. During compounding, the average melt temperature and the screw speed were set at 250°C at 85 rpm, respectively, resulting in a throughput of 1 kg/h using a gravimetric feeding. After extrusion, the blends were air-cooled and pelletized.

4.3.1.6 Bead Foams

Bead foams are produced by means of foam extrusion combined with underwater granulation. There are two approaches present for the fabrication of foamed beads – the creation of expandable beads and the production of already expanded beads. The first approach can be utilized for amorphous thermoplastic resins only. Here, polystyrene is taken as an example, since they retain a blowing agent in a solid state. Expandable beads are polymer granules where a blowing agent is trapped and the impregnated polymer granules are expanded in a separate step before sintering. Efficient transportation of the unfoamed blowing agent-impregnated polymer material and control of density are the main advantages compared to expanded beads [17]. Expanded beads are made from semi-crystalline thermoplastics, as the presence of crystalline domains restricts the storage of a blowing agent inside the solid bead [21]. An overview of the possible methods for producing polymer bead foams is given in Figure 4.2. The most often used method to produce high quantities of expandable beads of polystyrene is the suspension-polymerization with a blowing agent [22]. Two steps are present in the whole process, formation of the granules via polymerization

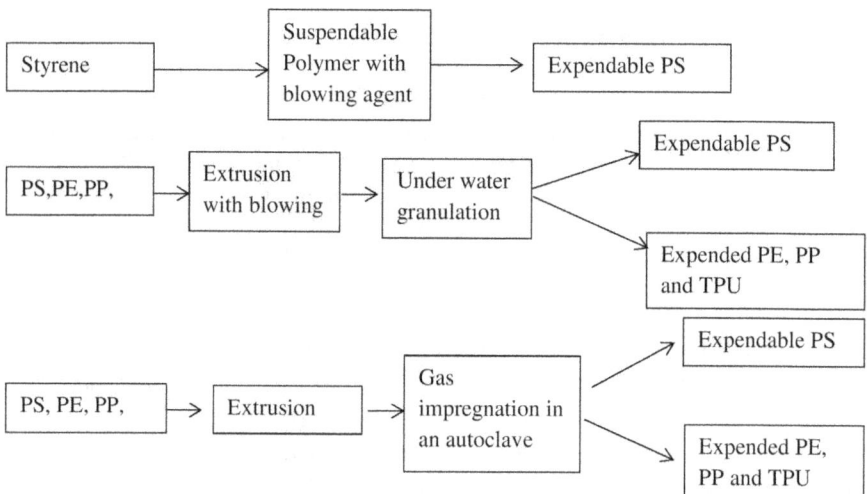

FIGURE 4.2   Different methods for the production of expandable and expanded beads.

and the addition of pentane and/or other blowing agents, which is being diffused into the granules [23]. On completion of this process, the beads were sieved to get several fractions with a narrow size distribution and were coated with antistatic agents to prevent agglomeration [24].

Another method to produce expanded beads is the impregnation of micro-granules along with the blowing agents, having all other required additives, in an autoclave. This is the main production process for EPP [23]. In the first impregnation vessel, the solid PP beads are saturated with gas close to the melting point of PP (150°C). After the saturation step, the materials along with the blowing agent are released to an expansion vessel [23]. Afterward, the beads are washed to remove any residual suspension stabilizer, which would inhibit proper welding of the beads [10]. For amorphous polymers, expandable beads can be produced as well, if the saturation step takes place at a temperature below the glass transition of the polymer-blowing agent solution. Alternatively, foam extrusion with underwater pelletizing allows the production of expandable beads or already expanded beads [24]. It is schematically shown in Figure 4.3. In this method, gas-loaded polymer melt is extruded through a hole plate into a water stream and cut by rotating knives. If the water pressure is above the vapor pressure of the blowing agent, the blowing agent is trapped within the solidifying polymer during cooling and expandable beads are produced. At low pressure, the dissolved gas evaporates and forms bubbles resulting in expanded beads. Advantages of this method are the exact dosing of the blowing agent(s) into the melt, a continuous and flexible process, and the applicability of additive that cannot be used in suspension polymerization [22,24], which theoretically allows the processing of any thermoplastic polymer with additives. Furthermore, the bead size is rather uniform [21]. The required temperature has to be kept constant throughout the experiment, otherwise, it may lead to unusable products [23]. Variable process parameters are the temperature and pressure of the water, the rotational speed of the knives, and the temperature of the perforated plate.

Recent Developments of New Bead Foams

Thermoplastic Elastomers and Thermoplastic Polyurethane Foams

Thermoplastic elastomers are unique materials with a wide range of properties, which lie between thermoplastics and elastomers. It's having the combined property of elastomers and the easy processability of thermoplastics. They differ according to structure, rheological, and thermal properties. Thermoplastic polyurethanes

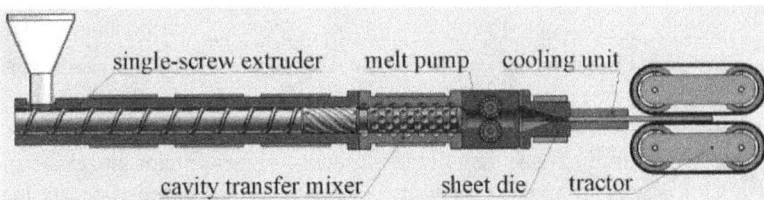

FIGURE 4.3 Schematic diagram of the single-screw extruder foaming system.

(TPUs) consist of a phase-separated molecular structure of rigid hard segment (HS) domains dispersed in a soft segment (SS) matrix, which provides TPUs with a unique combination of strength, flexibility, and processability. Irrespective of these facts, the application of TPUs is limited due to their high hardness and cost. Foaming became important since it could be a desirable way of reducing density and thereby cost. Expanded TPU (ETPU) bead foams and its product have been developed recently [25].

One example is the shoe sole of the Adidas "Energy Boost" running shoe, where ETPU with brand name "Infinergy" from BASF SE with relatively high foam densities of 200–300 g/l is used. The advantage is that softer TPUs have a lower concentration of HSs and hence a lower melting point with much better flowability. Thus, lower processing temperatures and pressures are required during the processing of ETPU beads. Furthermore, lower steam pressures are required during the sintering step of the ETPU beads, which reduces the sintering cost. The softness of the beads also makes adhesive bonding of the ETPU beads more effective.

The ETPU beads can be produced both in a batch setup using a high-pressure autoclave as well as a continuous setup using foam extrusion and an underwater pelletizer. During the processing of ETPU beads in a high-pressure autoclave setup, the TPU micropellets are fed into the high-pressure autoclave with suspension medium and saturated with the blowing agent. The autoclave is then heated to the impregnation temperature, which is generally near the softening point of the TPU. The blowing agent impregnates the SS and results in the swelling of the TPU. The impregnation time is generally from 0.5 to 10 h [25]. The impregnation with a blowing agent near the softening point of TPU results in the rearrangement of the existing HS crystalline domains. After completion of the saturation cycle, the autoclave chamber is depressurized, which results in the production of ETPU beads. During the depressurization process, cooling of the TPU material will take place and new HS crystallites are formed in the microstructure. The HS crystallites in the TPU microstructure can be effectively utilized as heterogeneous bubble nucleating agents to produce microcellular ETPU bead foams [26].

Polyester Foams
Foaming of polyesters — polyethylene terephthalate, PBT, or polylactide acid (PLA) is defying since they are not good in rheological properties for foaming due to its low melt elasticity and low viscosity [27]. It may lead to an unfavorable cellular morphology. They are mainly used for foam cores in sandwich applications. Extrusion foaming is the most used production method and allows the production of foams with densities between 30 and 400 kg/m^3 [28]. Recently, the bead-foaming capability of PBT was studied for expanded PBT using extrusion with an underwater pelletizing system [29]. The lowest density achieved for the processed EPBT beads was 280 kg/m^3. The effects of processing parameters like knife speed, water pressure, and temperature as well as the viscosity of the PBT grade were investigated. A variation of processing conditions showed moderate influence on bead shape and cellular morphology. Viscosity is an important parameter and hence increased viscosity leads to a better bead shape and cellular structure.

Biopolymer-based Foams
Among different biopolymers, polylactide acid (PLA) has gained a lot of interest for bead foaming applications. The interest is due to its high potential to substitute EPS bead foam products used in packaging and commodity products [30]. The production of EPLA beads has been commercialized and similar techniques are employed in the processing of EPS beads. Here the PLA micro-pellets are saturated with a blowing agent below PLA's glass transition temperature. The pre-foaming is done in a pre-expander machine, using steam or hot air. The expanded EPLA beads are then sintered together using a steam-chest molding machine. Prior to sintering, the pre-foamed PLA beads are coated with a special coating to improve the sintered EPLA beads for the steam-chest molding process. The thermal and mechanical properties of EPLA beads processed with the technique described above are similar to the EPS bead foam product compared at a similar density. The utilization of the double melting peak crystal formation technique is believed to be a promising method to improve the crystallization kinetics of PLA and also the sintering effects of EPLA beads. It is believed that the presence of dissolved gas will significantly improve the crystallization kinetics of PLA, and it will be beneficial for the production and sintering of EPLA beads [31,32]. The crystals that are formed during the processing of EPLA beads could also enhance the poor foaming behavior of PLA. The generated crystals during the saturation process will also promote cell nucleation by acting as heterogeneous nucleating agents. Furthermore, the expandability of EPLA is enhanced by the connection of PLA molecular chains through the generated crystals improving the low melt strength of PLA and consequently increasing beads' stability by minimizing the gas loss and cell coalescence.

Foamed Wood-Polymer Composites
In comparison with normal polymer compounds, the wood-polymer composites were found to be more advantageous in terms of material costs and stiffness [33]. But those advantages were obtained at the expense of ductility and impact strength. Another adverse effect observed was the increase in polymer density that can be a drawback in terms of application. Thus another methodology, for optimizing the different properties, was investigated, and thus came the foamed wood-polymer composites. The studies are in progress and the wood-foam products with the micro-porous structure are aiming to meet the properties such as high strength, lightweight components in automotive, high-speed rail, aerospace, military, and other fields by the virtue of cost-effective ratio and high-specific strength [34].

Nanoporous Polymer Foams
Polymer foam itself means the polymer with a porous structure. The pore size can be of different ranges. Initially, polymer foams with variable pore sizes were synthesized and the materials have been employed in various applications such as in packaging and building, aeronautics, automotive and shipbuilding industries, and wind turbines [35]. Also, many applications were found in different areas, viz. thermal insulation [36], acoustic insulation [37], scaffolds [38–40], and filtration [41], or catalysis processes because of their peculiar structure and high-specific surface

area [42]. These polymer foams were having pore sizes of around 100 μm. When the pore size was reduced to 10 μm, the polymer foams had shown increased mechanical properties such as compressive, flexural, and impact strength. These studies lead to the development of nanoporous polymer foams. It has been predicted that nanoporous polymer foams will be having better mechanical properties – toughness, higher impact energies, and strain to failure [43]. In addition to that, they may be having an improved thermal insulation behavior because of the Knudsen effect [44].

4.4 CHARACTERIZATION

Depending on the type of application, the blended polymer foams are characterized. Different characterization techniques are there, which are cited below

4.4.1 SWELLING STUDIES

The volume swelling ratio of the blended polymer foam in compressed CO_2 was measured from 0 to 25 MPa using a high-pressure cylindrical device with a transparent window [45]. The volume swelling ratio is the volume ratio of the polymer before and after the CO_2 sorption,

$$\text{Volume swelling ratio } (P, T) = V(P, T, wCO2) / V0(P, T) = h1 / h0$$

$V(P, T, wCO_2)$ is the volume of the polymer saturated with CO_2 at given pressure and temperature, $V_0(P, T)$ is its volume at the same pressure and temperature without CO_2, and h_0 and h_1 are the heights of the polymer before and after the CO_2 sorption.

4.4.2 CO_2 SOLUBILITY MEASUREMENT

The apparent solubilities of CO_2 in EPDM were measured directly by using MSB at CO_2 pressures from 5 to 25 MPa and at temperatures from 100 to 150°C. The details of the MSB apparatus and measurement procedures used can be found in reference [46].

4.4.3 SEM MEASUREMENT

The cell morphologies of the polymer foams can be visualized using a scanning electron microscope (SEM). The samples were immersed in liquid nitrogen for 3–5 min and then fractured. Before the SEM observation, the samples were gold-sputtered in a vacuum chamber. SEM images of the morphology of PBT bead foams are shown in Figure 4.4.

4.4.4 DSC MEASUREMENT

The melting behavior of the polymer (70/30) blend before and after foaming was determined using a differential scanning calorimeter calibrated with indium. During

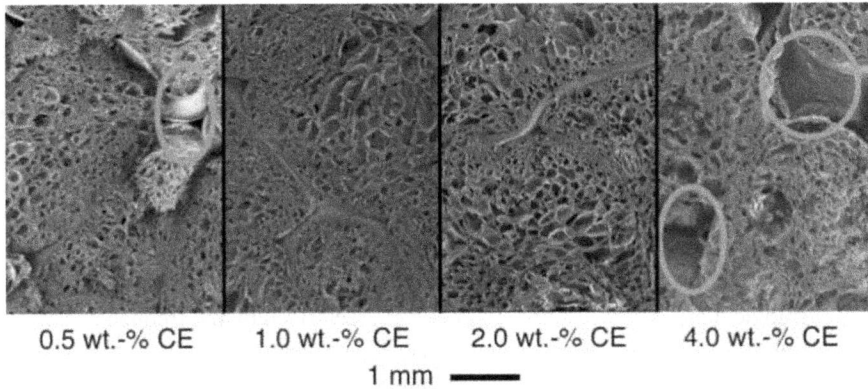

0.5 wt.-% CE   1.0 wt.-% CE   2.0 wt.-% CE   4.0 wt.-% CE

1 mm ━━━

FIGURE 4.4   SEM images of the morphology of the modified PBT bead foams [47].

the DSC tests, a thermal scanning from 50 to 200°C with a heating rate of 10°C/min was applied. The glass transition temperature (Tg) of the neat materials and the immiscible blend systems can be measured using DSC. The method consists of a heating–cooling–heating cycle under a nitrogen atmosphere from 25 to 250°C at a scanning rate of 10 K/min. The values of the second heating cycle were evaluated to calculate the Tg. Additionally, modulated DSC measurements at low temperatures were done with pure blends independently. This method gives information on the reversing and non-reversing characteristics of thermal events.

In order to understand the plasticizing effect of CO_2 under pressure on the glass transition temperature, the specimens were analyzed with a high-pressure DSC. The samples were saturated with CO_2 in the range of pressures (1, 10, 35 bar) for a minimum of 24 h. The measurement was performed from 25 to 260°C with a heating rate of 10 K/min under an inert atmosphere (N_2).

4.4.5 TENSILE AND COMPRESSION TESTING

Tensile tests of the unfoamed and foamed blended polymer (70/30) were performed on specimens of 4-mm width and 2-mm thickness at 23°C, according to ASTM D-638 at a cross-head speed of 20 mm/min. For the foams, the samples were cut into 2-mm thick sheets before testing. The compression tests of the foams were performed with an MTS universal tester microtester equipped with a 50 N load cell. The maximum strain is set to about 100% and the compression rate is set at 0.5 mm/min. The samples are compressed several times.

4.4.6 MORPHOLOGICAL CHARACTERIZATION

A transmission electron microscope was used to determine the morphology of the blends after extrusion. The ultrathin cuts were prepared with a Reichert–Jung ultramicrotome (Ultracut E) and then stained with osmium tetroxide (OsO4) for 2 h at ambient pressure.

4.4.7 Density Measurements

To measure the apparent density of each sample, the weight and volume of the sample were separately determined using an electronic balance with an accuracy of 0.01 g and a micrometer caliper with an accuracy of 0.01 mm. The weight of the sample was divided by the sample's volume to obtain the apparent density. Besides, five replicates were used for each sample and the mean value was taken.

4.4.8 Rheological Property Testing

The rheological properties of samples can be characterized by the frequency dependencies of molten-state oscillatory shear moduli, such as the shear storage modulus G' and the loss modulus G'', which were conducted using a parallel-plate rheometer. The tests can be performed in dynamic mode between parallel plates (diameter 25 mm, gap 1 mm) at 180°C in a nitrogen atmosphere. Besides, in the process of testing, frequency sweeps plates (diameter 25 mm, gap 1 mm) at 180°C in a nitrogen atmosphere.

4.5 CONCLUSION

It can be stated that blended polymer foams are having additional factors determining the foamability of multiphase blends, besides the commonly accepted key criteria of neat polymers, including sufficient gas solubility and diffusion rates under suitable processing conditions and balanced rheological properties. In the case of multiphase blends, the significant influence of the initial blend morphology, of the rheological, as well as of the thermal properties of the respective phases and the blend needs, to be taken into account. The refined foam morphology allows to enhance the mechanical properties of foamed blends when compared to neat polymers. However, the exploitation of the blending approach appears even more promising to establish cellular barrier materials.

Wide ranges of applications are now identified as the application of polymer blend foams. Even the natural rubber is foamed and is applied in the tire-making process. The foamed rubber is being used for the shock absorbance of vehicles. Also, the characterization process is not much costly, and simple methods that are economically viable can be employed for characterization. Hence in the future, much more applications will be developed and thus polymer blend foams are going to be well acclaimed and utilized.

REFERENCES

1. Holger R, Volker A, Axel HEM. *Foaming of polymer blends—chance and challenge, anais do 9o Congresso Brasileiro de Polímeros*, Proceedings. https://doi.org/10.1177/026248930702600602
2. Stefanie B, Ronak B, Tina IL, Holger S, Axel HEM, Volker A. Polymer foams made of immiscible polymer blends compatibilized by janus particles—effect of compatibilization on foam morphology. *Advanced Engineering Materials* 18(15) (2015) 814–825.

3. Pinto J, Dumon M, Rodriguez-Perez MA. *Recent developments in polymer macro, micro and nano blends*, Woodhead Publishing 2016, Book Chapter 2017.
4. Raps D, Hossieny N, Park CB, Altstadt V. Past and present developments in polymer bead foams and bead foaming technology. *Polymer* 56 (2015) 5–19.
5. Luo Y, Xin C, Yang Z, Yan B, Li Z, Li X, He Y. Solid-state foaming of isotactic polypropylene and its composites with spherical or fibrous poly(butylenes terephthalate). *J. Appl. Polym. Sci.* 2015 (2015) 132.
6. Rizvi A, Tabatabaei A, Barzegari MR. In situ fibrillation of CO_2-philic polymers: Sustainable route to polymer foams in a continuous process. *Polymer* 54 (2013), 4645–4652.
7. Ali R, Chu RKM, Lee JH, Park CB. Superhydrophobic and oleophilic open-cell foams from fibrillar blends of polypropylene and polytetrafluoroethylene. *ACS Appl. Mater. Interfaces* 6 (2014) 21131–21140.
8. Boonprasertpoh A, Pentrakoon D, Junkasem J. Effect of crosslinking agent and branching agent on morphological and physical properties of poly(butylene succinate) foams. *Cell Polym.* 36 (2017) 333–354.
9. Feng Z, Luo Y, Hong Y, et al. Preparation of enhanced poly(butylene succinate) foams. *Polym. Eng. Sci.* 56 (2016) 1275–1282.
10. Li G, Qi R, Lu J, et al. Rheological properties and foam preparation of biodegradable poly(butylene succinate). *J. Appl. Polym. Sci.* 127 (2013) 3586–3594.
11. Lim SK, Lee SI, Jang SG, et al. Fabrication and physical characterization of biodegradable poly(butylene succinate)/carbon nanofiber nanocomposite foams. *J. Macromol. Sci. B* 50 (2010) 100–110.
12. Lim S-K, Jang S-G, Lee S-I, et al. Preparation and characterization of biodegradable poly(butylene succinate)(PBS) foams. *Macromol. Res.* 16 (2008) 218–223.
13. Lee EK. *Novel manufacturing processes for polymer bead foams*. University of Toronto; 2010.
14. Doroudiani S, Park CB, Kortschot MT. Effect of the crystallinity and morphology on the microcellular foam structure of semicrystalline polymers. *Polym. Eng. Sci.* 36 (1996) 2645–2662.
15. Britton R. Update on mouldable particle foam technology. *iSmithers*; 2009.
16. Scheirs J, Priddy D, editors. *Modern styrenic polymers*. John Wiley & Sons Ltd; 2003.
17. Colton JS, Suh NP. Nucleation of microcellular foam: theory and practice. *Polym. Eng. Sci.* 27 (1987) 500e3.
18. Lee ST, Ramesh NS. *Polymeric foams e mechanisms and materials*. CRC Press; 2004.
19. Altstadt V, Mantey A. *Thermoplast e Schaumspritzgießen*. Hanser Verlag; 2010.
20. Britton R. Update on mouldable particle foam technology. *iSmithers*; 2009.
21. Scheirs J, Priddy D, editors. *Modern styrenic polymers*. John Wiley & Sons Ltd; 2003.
22. Wagner J. *Halogenfreie Flammschutzmittelmischungen für PolystyrolSch€ aume*. Ruprecht-Karls-Universitat Heidelberg; 2012.
23. Eyerer P, Hirth T, Elsner P. *Polym. Eng.*, Technologien und Praxis. Springer Verlag, Berlin, 2008.
24. Koppl T. *Halogenfrei € flammgeschütztes Polybutylenterephthalat und dessen Verarbeitung zu Polymerschaumen*. University of Bayreuth; 2014.
25. Prissok F, Braun F. *US 20100222442 A1-foams based on thermoplastic polyurethanes*, 2010.
26. Hossieny NJ, Barzegari MR, Nofar M, Mahmood SH, Park CB. Crystallization of hard segment domains with the presence of butane for microcellular thermoplastic polyurethane foams. *Polymer* 55 (2014) 651–662.
27. Raps D, Heymann L, K€oppl T, Altst€adt V. Effect of supercritical CO_2, pressure and flame retardant on rheological properties of PBT in shear deformation. PPS 29-B. *Abstr.* 2013.

28. Borer C, Naef UG, Culbert BA, Innerebner F. Mechanical Properties of Strand PET Foams at Different Length Scales. *Dissertation 19800166*, 1999.

29. Strasser JP. *WO 2011063806 A1-process for producing pet pellets, and pet pellets*, 2011.

30. Koppl T, Raps D, Altst€adt V. E-PBT-Bead foaming of poly(butylene terephthalate) by underwater pelletizing. *J. Cell Plast.* 50 (2014) 475–487.

31. Garlotta DA. Literature review of poly (lactic acid). *J. Polym. Environ.* 9 (2001) 63–84.

32. Nofar M, Zhu W, Park CB. Effect of dissolved CO_2 on the crystallization behavior of linear and branched PLA. *Polymer* 53 (2012) 3341–3353.

33. Zhang S, Rodrigue D, Riedl B. Preparation and morphology of polypropylene/wood flour composite foams via extrusion. *Polym. Compos.* 26 (2010), 731–738.

34. Suwei W, Ping X, Mingyin J, Jing T, Run Z. Effect of polymer blends on the properties of foamed wood-polymer composites. *Materials* 12 (2019), 1971.

35. Solórzano E, Rodriguez-Perez MA. Polymer foams, in: M. Busse et al. (Ed.), *Structural Materials and Processes in Transportation*, Wiley-VCH, Weinheim, Germany, 2013.

36. Papadopoulos AM. State of the art in thermal insulation materials and aims for future developments. *Energy Build.* 37 (1) (2005) 77–86.

37. Díez-Gutiérrez S, et al. Technical note. Impact sound reduction of crosslinked and non-crosslinked polyethylene foams in suspended concrete floor structures. *J. Build. Acoust.* 10 (3) (2003) 261–271.

38. Ghosh S, et al. Dynamic mechanical behavior of starch-based scaffolds in dry and physiologically simulated conditions: effect of porosity and pore size. *Acta Biomater.* 4 (4) (2008) 950–959.

39. Prabaharan M, et al. Preparation and characterization of poly(L-lactic acid)-chitosan hybrid scaffolds with drug release capability. *J. Biomed. Mater. Res. Part B, Appl. Biomater.* 81 (2) (2007) 427–434.

40. Hentze HP, Antonietti M. Porous polymers and resins for biotechnological and biomedical applications, *Rev. Mol. Biotechnol.* 90 (1) (2002) 27–53.

41. Ulbricht M. Advanced functional polymer membranes. *Polymer* 47 (7) (2006) 2217–2262.

42. Twigg MV, Richardson JT. Theory and applications of ceramic foam catalysts. *Chem. Eng. Res. Design* 80 (2) (2002) 183–189.

43. Miller D, Kumar V. Microcellular and nanocellular solid-state polyetherimide (PEI) foams using sub-critical carbon dioxide II. Tensile and impact properties. *Polymer* 52 (13) (2011) 2910–2919.

44. Lu X, et al. Correlation between structure and thermal conductivity of organic aerogels. *J. Non-Crystall. Solids* 188 (3) (1995) 226–234.

45. Chen J, Liu T, Zhao L, Yuan WK. Experimental measurements and modeling of solubility and diffusivity of CO2 in polypropylene/micro- and nanocalcium carbonate composites. *Ind. Eng. Chem. Res.* 52 (2013) 5100–5110.

46. Areerat S, Funami E, Hayata Y, Nakagawa D, Ohshima M. Measurement and prediction of diffusion coefficients of supercritical CO_2 in molten polymers. *Polym. Eng. Sci.* 44 (2004) 1915–1924.

47. Tobias S, Bianca H, Michael F, Volker A. Development of a Bead Foam based on the Engineering Polymer Polybutylene Terephthalate, AIP Conference Proceedings 2055, 060004 (2019).

5 Polymer Nanocomposite Foams as Metal Ion Removers

Arvil Dasgupta, Souhardya Bera, Arnab Roy,
Rishov Kumar Das and Subhasis Roy

CONTENTS

DOI: 10.1201/9781003218692-5

5.1   INTRODUCTION

An environment is made up of the earth's plants, water, and atmosphere, which influence humans, microorganisms, plants, and animal life. Among so many spheres, the biosphere is the most important as it accommodates the living organisms which interact with their non-living soil, air, and water. In the late centuries, globalization had launched such components that diminished the integrated functioning of the environment and its ability to nurture life [1]. Nowadays, the atmosphere is being polluted and contaminated. Pollution is the introduction of substances and energy by humans into nature that has harmful effects on living resources and can impair the environment's physical-chemical and biological quality. On the other hand, contamination is the raised level of substances in the area and the organisms [2].

Industrial, agricultural, and domestic wastes cause hazardous pollution, endangering living beings worldwide. Among the multiple water pollutants, heavy metals do not biodegrade and tend to have the capacity to concentrate in living beings, which causes a severe impact on human health and initiates ecological damage. Agro-industrial processing, textile dyeing, and tanneries continuously produce toxic pollutants and drain in ponds, lakes, rivers, etc. This causes various metal ions pollution in water such as Pb^{2+}, Hg^{2+}, Cd^{2+}, Zn^{2+}, Cu^{2+}, Cr^{3+}, etc. The increased concentration of these metal ions is the potential parameter that causes critical health issues, such as esophageal ulcers, gastric issues, neural damage, and liver diseases. Thus, we need to take the proper measures to control the concentrations of the metal ions in water. Although techniques such as ion exchange, reverse osmosis, chemical precipitation, and electrochemical treatment exist to reduce metal ion concentration in wastewater, processes need to be more efficient and cost-efficient [3]. In this case, polymeric foams catch massive importance in their application in metal ion reduction. This chapter depicted the nature of various types of polymeric foams and their utilities for reducing multiple metal ions in water.

5.2   POLYMERIC NANOCOMPOSITE FOAMS, THEIR TYPES AND APPLICATIONS

Polymeric foam has vast applications, and they are found almost everywhere in our modern world. One of the most critical applications is its usage to remove metal ions. Polymeric foams have become very popular due to their various properties, like lightweight, thermal insulation, acoustic insulation, and energy absorption.

5.2.1 Polymeric Nanocomposite Foam

Polymer nanocomposite foams are defined as polymer or copolymers having nanoparticles or nanofillers dispersed in the polymer matrix and are less than 100 nm in either of its dimensions at minimum. Polymeric Nanocomposite foam is a kind of gaseous phase dispersion tentatively in a solid-phase polymer environment by mixing different nanoparticles like nanoclays, carbon nanotubes, and graphene with

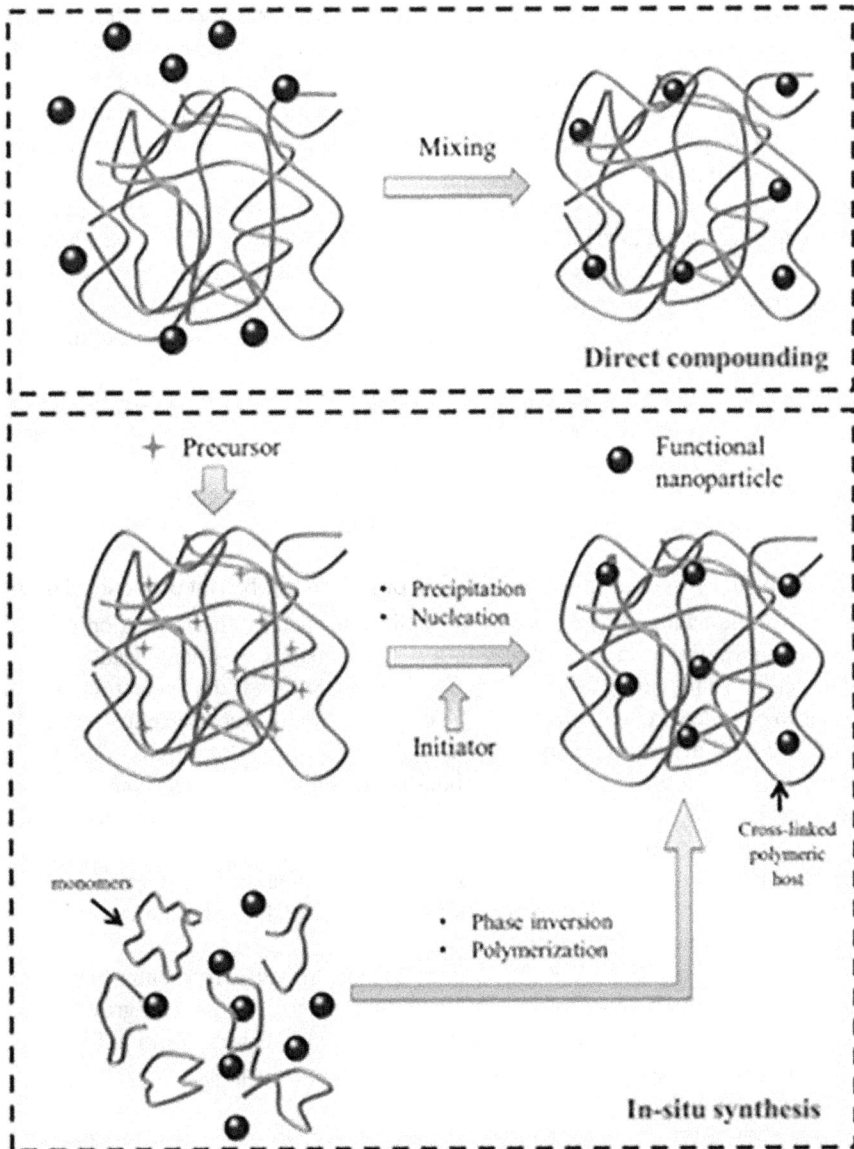

FIGURE 5.1 Illustration of synthesis methods for PNCs [2] (Reprinted with permission).

in-situ polymerization. The structure, size, and surface chemistry of nanofillers are often modified to regulate the froth structure and, thus, the properties of the foam. Nanoparticles also add functionality to polymer foams [1]. During the polymerization of monomers, an additional *in-situ* synthesis method was to add functional nanoparticles (Figure 5.1).

Generally, the gaseous phase does the expansion (acts as a blowing agent), and the solid phase builds up the polymer matrix. There may be the presence of other solid phases that act as fillers. For the formation to occur, the two phases of the foam are rapidly combined. The distinction can be vividly understood between the foam of polystyrene and polyurethane, which are closed foam and open-cell foam, respectively [1,3]. Polymeric foams are further divided into thermoplastics and thermosetting. Polymeric foams can be modified by adding nanoclays, which are nano-sized particles by the process of melt compounding, or *in-situ* polymerization. It is possible to develop partial dispersion of nanoclay distribution (intercalated) or full nanoclay (exfoliated) by using various surfactants to modify the particle surface and control the processing conditions [3]. Supercritical carbon dioxide is a low-cost and nonflammable gas, it is commonly utilized as a blowing agent. The thermoplastics foams can be broken down quickly by applying heat and recycling. In contrast, thermosetting foams are hard to recycle because heavily cross-linked polymers are involved in the mechanism [4].

Polymeric nanocomposite (PNC) foams are so widely used mainly because of the advantageous properties that they hold. The density is low, and the weight of these foams is comparatively low. Some polymeric foams have deficient heat transfer properties, which can also be used as insulators. They have excellent absorption properties, due to which metal ions can easily be removed.

5.2.1.1 Basic Principles in the Formation of Polymeric Nanocomposite Foams

The polymeric nanocomposite foams are mainly formed by the nucleation process and the incorporation of gas bubbles in a polymer matrix. A micro-bead of gas is entrapped into a polymer system or latex in synthetic foams. If the nucleation mechanism is to be followed, with growing bubble size, the foam structure changes through several stages. Varieties of nanofillers may be used to alter the strengths of the nanocomposite foam. It can be seen that carbon-based nanofillers like carbon nanotubes and graphene may enhance polymeric foams' mechanical strength and conductivity [5]. Various surface properties also need to be considered, e.g., a better nucleating surface is provided by the flat surface of graphene compared to the curved surfaces of the carbon nanotubes [6]. It should then be considered that carbon nanotubes provide several nucleation sites along their length.

Firstly, during the formation of polymeric foams, tiny dispersed bubbles are produced in a continuous phase of the liquid polymer matrix. Further cell growth leads to lower foam density. Viscosity and surface tension also cause to form uniform cell structure. Due to the Cooling of closed-cell formation, shrinkage of the polymeric foam occurs for reduced pressure inside the cell.

Formation of polymeric nanocomposite foams through mechanical, chemical, or physical methods:

i. In the case of the Thermal method, the decomposition of chemical blowing agents takes place, which leads to the generation of carbon dioxide or nitrogen or both, due to the application of heat or the exothermic reaction during polymerization. Chemical blowing agents used in this case are either organic materials like azides, hydrazides, nitroso compounds, or inorganic materials like bicarbonates, carbonates and borohydrides [7].

ii. The mechanical method is implemented by generating flotation of gases into a polymeric fluid system that can either be a solution or a suspension. Catalytic action, heat, or both combined, can get the polymer hardened, and hence gas bubbles can get entrapped in the polymer matrix [3].

iii. The chemical method is done by using a blowing agent in the *in-situ* reaction during polymerization, e.g., the reaction of water with isocyanate, leading to the formation of carbon dioxide, which acts as the blowing agent and is responsible for the formation of polyurethane foam. Expansion of dissolved gas in a polymer mass at reduced pressure and absorption of microspheres or nanoparticles are two more methods that can be used [3].

5.2.2 Types of Polymeric Nanocomposite Foams to Remove Metal Ions

5.2.2.1 Magnetic Polyurethane Nanocomposite Foam

Magnetic Fe_3O_4 nanoparticles were modified with (3-AminoPropyl) triethoxysilane (APTES) by an *in situ* polymerization process and then can be used in various weight percentages (3, 6, 8, 12, and 20 wt%) into a polyurethane foam matrix [8]. This mainly uses the removal of arsenic and Cd (II) heavy metals from drinking water.

5.2.2.2 Carboxy-methylated Cellulose Nanofibrils (CMCNFs) Embedded in Polyurethane Foam

The composite foams are formed from dispersed Carboxy-methylated Cellulose Nanofibrils (CMCNF) in polyurethane foam matrices having an open-pore structure. The presence of hydrogen bonds enriches mechanical strength. It is used as an adsorbent for wastewater treatment to treat heavy metal ions. The composite foam has a high adsorption capacity and a high recyclability potential. Polyurethane reacts with other compounds, including isocyanate-reactive moieties such as hydroxylamine and carboxyl functional groups, to produce hydrophilic PU polymer. The reactive component is often an aqueous phase, and the ratio of PU polymer to an aqueous phase can be adjusted to produce hydrophilic foam with various cellular architectures [9,10]. This mainly uses the removal of Cu^{2+}, Cd^{2+}, and Pb^{2+} heavy ions.

5.2.2.3 Graphene Reinforced Nanocomposite Foams

Polyacrylonitrile (PAN) belongs to the class of thermoplastic acrylic polymers. It possesses high thermal stability and strength. PAN exists in various forms, including film, foam, fiber, membrane. PAN also contains the potential to soak up toxic metal ions. Poly (vinylidene fluoride) (PVDF) also belongs to the class of thermoplastic polymers, having good heat conductivity, chemical defiance, and oxidation resistance. Organic toxins and metal ions are usually removed using PVDF-based membranes. Graphene is a highly robust nanocarbon material with great elastic strength and high thermal conductivity. Graphene Nano Platelet (GNP), is a high-performance nanocarbon that can be used as an intelligent reinforcing nanofiller for polymeric matrices. The PAN/Poly(VinyliDeneFluoride-co-exaFluoroPropylene)/ GrapheneNanoPlatelet (PAN/PVDF-HFP/GNP) blend is mainly prepared by dissolving PAN in DMF, and continuously stirring for 24 h at 50°C before adding PVDF in that solution. GNP nanofillers (0.5–5% by weight) are added to the blended mixture after it has been refluxed at 130°C for 6 h. A reflux time of 6 h is allowed for the reaction mixture at 120°C. Then the foaming agent is added, and it's further refluxed [11]. This mainly uses the removal of Pb^{2+} and Ni^{2+} ions.

5.2.2.4 Elastomeric Nanocomposite Foams

Modifications are done on the Poly Di-Methyl Siloxane (PDMS) foams, such that they become hydrophilic and allow the contaminated water to pass through their volume. ZnSe colloidal nanocrystals (NCs), are placed on the foam surface to entrap the metal ions. The hydrophilic PDMS foams were developed by applying the leaching technique [11]. These nanocomposite foams are used in polluted water or wastewater treatment to remove Pb^{2+} or Hg^{2+} with high efficiency by cation exchange reaction mechanism.

5.3 MECHANISM OF REMOVAL OF METAL IONS

The primary mechanism of the PNC foam removing the metal ions is adsorption. Adsorption occurs when the solute molecules (adsorbate) are selectively retained on a solid (adsorbent) surface. A specific type of nanocomposite polymeric foam is selected based on the metal ions. Many criteria need to be considered while choosing an adsorbent [12].

5.3.1 SELECTIVITY

The adsorbent can retain the adsorbate, and not all the molecules are present selectively. The selectivity depends on the following factors.

5.3.1.1 Affinity
When the target molecules have a higher affinity to the adsorbent than the other molecules present, adsorption occurs. The higher the affinity, the better the selectivity.

5.3.1.2 Intraparticle Diffusion Rate
The adsorbents are generally porous in structure, so different molecules have different diffusion rates through the adsorbent molecule. The selectivity due to the difference in diffusion rate is also known as kinetic selectivity.

5.3.1.3 Size
In many cases, the selectivity is based on the size of the adsorbate. If the target molecule has a larger size than the pore size, then the molecule may not enter the

pore, and those having a size lesser than the pore will be adsorbed. According to the system, selectivity may function all the factors.

5.3.2 ADSORPTION CAPACITY

An adsorbent can adsorb. An adsorbent having high adsorption capacity is chosen so that the amount of adsorbent to be used is low. It is dependent on the affinity and the specific surface area of the adsorbent.

5.3.3 REVERSIBILITY OF ADSORPTION

It is the ability of an adsorbent to be used again in a new batch after being used for adsorption.

So, the PNC foam is selected, keeping all these factors in mind.

5.3.4 ADSORPTION ISOTHERM

Adsorption isotherm is the equilibrium relationship between the amount of solute adsorbed per unit mass of the adsorbent and the concentration of the adsorbate or solute at a constant temperature. Many adsorption isotherms are being generated using a reasonable theoretical basis, and some are purely empirical [12].

5.3.4.1 Langmuir Isotherm

The main principle behind the Langmuir isotherm is physical adsorption. It follows certain assumptions: (1) molecules can only be adsorbed at particular discrete active sites on the surface, (2) a single active site can adsorb a single molecule only, (3) the adsorbing surface is energetically uniform, (4) adsorbed molecules do not have any interaction.

$$q_e = \frac{q_{mb}C_e}{1 + bC_e} \tag{5.1}$$

Here, the q_e (mg/g) is the amount of adsorbate adsorbed at equilibrium on the surface of the adsorbent.

C_e (mg/g) is the equilibrium concentration of the solution.

q_m (mg/g) and b (L/mg) are the two adjustable Langmuir parameters [12].

5.3.4.2 Freundlich Isotherm

One of the essential isotherms in the class of empirical isotherm, Freundlich assumes the amount of adsorbate adsorbed has a power-law dependence on the concentration, which is usually expressed by:

$$q_e = K_f C_e^{1/n} \tag{5.2}$$

q_e (mg/g) is the equilibrium adsorption capacity.

C_e is the equilibrium concentration of the adsorbate in the solution.

K_f (mg/g) and n are the Freundlich parameters [12].

5.4   CASE STUDY FOR VARIOUS METAL IONS

5.4.1 Pb^{2+} Lead (II), Hg^{2+} Mercury (II) Ion Removal

The average concentration of Pb^{2+} and Hg^{2+} ions must be below 1 and 10 ppb, respectively. Due to industrial growth and modernization, contamination sources for local water resources are also increasing. Most of them are related to cardiovascular and gastrointestinal diseases, besides having decreased kidney function, nervous system damage, and many other complications [11]. The removal efficiency of Hg^{2+} and Pb^{2+} ions from wastewater can be tested by various techniques, with the majors being adsorption, photocatalytic degradation, chemical precipitation, membrane filtration.

In some cases, weak interactions with metallic ions result in limited adsorption capabilities, and heavy metal removal becomes inefficient due to additional reasons, such as low chemical affinity. A way to increase the removal efficiency of these metals from water is by utilizing a three-dimensional polymeric network. Due to the nano and microporosity, high surface area, damping properties, and increased strength, these polymeric structures achieved due to the addition of nanofillers, metal ion removal become efficient by adsorption or ion exchange. Although various experiments for using polymer composite foams have been conducted, with a few of them based on polyurethane-based foaming systems for the removal of heavy metal ions such as lead (II) from aqueous solutions, the difficulty encountered in that technique is the presence of high filler concentrations [11,13].

In an experiment, Chavan *et al.* [11], prepared a novel elastomeric polymer nanocomposite foam. It is a modified hydrophilic PolyDiMethylSiloxane (PDMS) foam that is equipped with ZnSe NCs with aerogels or xerogels. Pb^{2+} and Hg^{2+} metal ion removal from contaminated water find the primary usage with even high efficiency. An ion-exchange mechanism is used to remove metal ions in most cases. The hydrophilic portion of the PDMS foam allows contaminated water to pass through it, while the ZnSe colloidal NCs that surround the foam's surface work as an active component to trap metal ions. Nanocomposite foams remove Pb^{2+} and Hg^{2+} from contaminated water by this method after being immersed in untreated wastewater. A cation exchange reaction has been discovered that the elimination of metal ions has a 97–99% efficiency [11,13].

5.4.1.1   Preparation Technique

PDMS is silicone-based elastomeric foam consisting of two parts – the base and a curing agent. For the fabrication of PDMS foams, generally, poly(ethylene oxide)-co-polydimethylsiloxane (PEO-b-PDMS), toluene and lead nitrate, and Mercury bromide are used.

The leaching procedure is used to create PDMS foams. Initially, at ambient temperature, the (PEO-b-PDMS) surfactant additives are combined with the PDMS polymeric foam base with continuous stirring. The hardening agent is added with a weight ratio of 0.1:10:1, respectively. Under low-pressure circumstances, the developed mixture is poured into a petri-dish with added sugar cubes. The solution perforates via the pores and fills the gaps of the sugar cubes due to the capillary action. The system for curing the pre-polymer is then placed in the oven at 80°C for a residence time of 24 h. Eventually, hot water is used to dissolve the sugar, followed by ultrasonication, which develops porous hydrophilic PDMS foams [14].

ZnSe NCs are synthesized from Zinc Stearate, octadecylamine, 1-octadecene, oleylamine, and selenium powder. Toluene and methanol are used for washing the nanocrystal. In the final stage, the prepared foam is modified with ZnSe NCs by dipping the NCs in the PDMS foams using a dip adsorption procedure. With ZnSe NCs at the surface, the foam is dipped in toluene solution and left at room temperature for 24 h before being dried under a nitrogen atmosphere where it regains its original dimension. This aids in the adsorption of a specified amount of ZnSe NCs in the foam, resulting in Elastomeric Nanocomposite foam [13,14]. This technique was adopted by Chavan *et al.* for making elastomeric foam [11].

5.4.1.2 Ion Exchange Mechanism Study

Scientists Chavan *et al.* [11] Conducted a series of experiments for the ion exchange mechanism study to remove Pb^{2+} and Hg^{2+} from the diluted solution to determine the removal efficiency. Firstly, 100 mL aqueous solutions containing 20 and 40 parts per million (ppm) of Pb^{2+} and Hg^{2+} were prepared. The ZnSe-loaded PDMS foams were introduced at ambient temperature in 10 mL of each metal ions solution with adequate stirring. After proper time intervals, the samples were taken out and evaluated. After a brief time interval, the color of the foams changed dramatically from yellow to brownish. When it was kept for 24 h, the foam became dark brown. Figure 5.2 shows elastomeric foam before, after the ZnSe loading, and after removing Pb^{2+} ions from water.

After the specified time interval, an Inductive Coupled Plasma-Atomic Emission Spectrometer (ICP-AES) was used to reveal the concentration of ions. The collected solutions were absorbed in aqua regia [10]. The following equation calculates the removal percentage of ions overtime—

$$\text{Removal of Ions } (\%) = \frac{q_0 - q_t}{q_0} \times 100 \tag{5.3}$$

Where,

q_0 is the initial concentration of ions (in ppm).
q_t is their concentration at a given time t (hour), (in ppm).

EDFS analysis and XRD analysis are generally applied to study the surface morphology after the ZnSe-PDMS foam is dipped in the Pb^{2+} solution. The XRD analysis mainly confirms the presence of PbSe NCs instead of ZnSe from the different peak patterns obtained at different 2θ values corresponding to further reflections from the planes of PbSe NCs after the cation exchange [14].

FIGURE 5.2 Elastomeric foam before (white), after the ZnSe loading (yellow), and after the removal of Pb^{2+} ions from water (dark brown) [11] (Reprinted with permission).

The modified PDMS foams are hydrophilic, allowing the contaminated aqueous solution to infiltrate. ZnSe colloidal NCs on the foam surface serve primarily as an active component to capture the metal ion. The primary cation exchange process between Zn^{2+} and Pb^{2+} and Hg^{2+} occurs after the nanocomposite foam is immersed in an aqueous solution of Pb^{2+} or Hg^{2+}. As a result, the heavy metal ions are successfully eliminated, and side products such as ZnO are generated due to the interaction of the liberated Zn^{2+} ions with oxygen atoms from water or the polymer [14].

5.4.1.3 Kinetics Study

When the ZnSe PDMS foams were immersed in a Pb^{2+} ion water solution, the variation of both ions' concentration was analyzed to study the interaction between Zn^{2+} and Pb^{2+} ions and to verify the Pb^{2+} ion removal. A ZnSe–PDMS foams (3 wt%) solution was immersed in two different samples containing Pb^{2+} ion in water solutions of 20.06 and 37.86 ppm, respectively. An instant drop of Pb^{2+} ions concentration in the first hour was observed (95.6% of the Pb^{2+} ions were removed) along with a significant increase in the concentration of the Zn^{2+} ion. The time of such changes was noted. With a 24-h residence time, the final removal percentage of 98.0% and

99.9% were observed in both low and highly concentrated Pb^{2+} ions solutions, with an allowed residence time of 24 h. As the results concluded, the kinetics of ions removal did not depend on the initial ion concentration.

The distribution coefficient (K_d) can estimate the affinity of the ZnSe for the Pb^{2+} metal ions by the following equation

$$K_d = (V/m)\frac{[C_0 - C_f]}{C_f} \qquad (5.4)$$

C_0 and C_f are the initial and final metal ion concentrations (ppm) in the water.
V = volume of the testing solution (mL).
m = the amount of solid exchanger (g) used in the experiment.

The value of K_d determines the quality. With a K_d value of 500 mL/g, the interaction with the metal ions is acceptable; a value above 5,000 mL/g is outstanding, with a value above 50,000 mL/g is termed outstanding. Figure 5.3 shows the kinetics of ion removal for both the concentrations

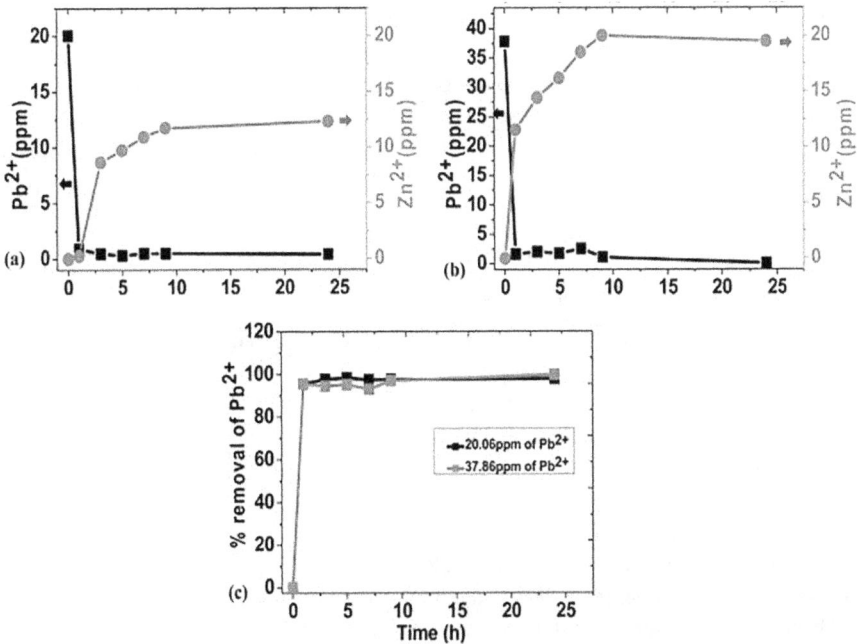

FIGURE 5.3 The figure shows the kinetics of ion removal at (a) 20.06 ppm of Pb^{2+} and at (b) 37.86 ppm of Pb2+ ion concentration. The third part shows (c) the removal percentage of Pb2+ ions for both concentrations. (A concentration of 3 wt% of ZnSe-PDMS foam was used.) [11] (Reprinted with permission).

In the experiment mentioned, at a Pb^{2+} concentration of 20.06 ppm in water, the K_d value was 2,102.46 mL/g, from which it can be concluded that the interaction of Pb^{2+} was in the acceptable range. On the other hand, at a value of 37.86 ppm of Pb^{2+} ions, the K_d value is pretty much close to the most efficient results, clocking a value of 42,371.34 mL/g, making it pretty much an exciting process [11,14].

To check the removal of the Hg^{2+} ions from water, the ZnSe–PDMS foams were tested. Even for a low concentration of the ZnSe NCs in PDMS foams (1 wt%), the removal percentage of Hg^{2+} ions was 99.5% within the first hour of reaction by the experiments conducted by Scientists Chavan et al. [11]. In a similar manner as in the case of Pb^{2+} ions, the removal of Hg^{2+} ions was also checked for two different concentrations of Hg^{2+} ions in water, i.e., 19.37 and 39.90 ppm. Precisely, the concentration of Hg^{2+} ions was reported to decrease to 0.0863 and 0.0607 ppm after 1 h of exposure, and their final concentration after 24 h was found to be 0.04 and 0.03 ppm for initial concentrations of 19.37 and 39.90 ppm, respectively [11]. This kinetic study indicates the high removal efficiency of Hg^{2+} ions from the polluted aqueous solution (above 99.8%) and shows outstanding K_d values using equation 4 (the obtained K_d values for the lower and higher concentration are 22,183.195 mL/g and 55,072.53 mL/g, respectively) [11,15]. There was also a sharp rise of Zn^{2+} concentration in water, confirming the ion exchange mechanism. The ability of the ZnSe-PDMS foam to remove Hg^{2+} has also been tested under the presence of other metal ions in the wastewater, e.g., Al^{3+}, Mg^{2+}, Fe^{3+}, Co^{2+}, etc., and from the obtained results it can be predicted that other metal ions do not affect the Hg^{2+} removal efficiency of the foam, which was 99.7% after 24 h, but the removal rate occurs more gradually, i.e., 30% in 1 h [11,16].

This observation demonstrates the adaptability of the elastomeric nanocomposite foams like ZnSe-PDMS foams for their accurate removal of different heavy metal ions under diverse conditions.

5.4.2 As^{3+}ARSENIC (III) ION REMOVAL

The maximum amount of arsenic in drinking water has been designated not to exceed 10 ppb. However, the arsenic concentration in many countries is above the limiting value. It has been reported that parts of Asia are highly affected by more arsenic concentration in the drinking water, followed by Europe, Africa, North America, South America, and Australia. The arsenic concentration in countries like Bangladesh, China, Nigeria, Ghana is above 50 ppb [17]. In some cases, water containing arsenic ions beyond the limiting concentration value is consumed over a long period. In that case, it will lead to devastating health-related issues like skin cancer, lung, kidney, and heart diseases. Therefore, it is a necessity to remove arsenic ions [18].

Hussein et al. demonstrated magnetic polyurethane foam nanocomposite to remove arsenic ions. Open-cell polyurethane foam's metal removal capacity was enhanced by doping it with magnetic Fe_3O_4 nanoparticles at various concentrations (4, 8, 12, 16) weight percentages. Their experiment discovered that the nanocomposite with 12 wt% magnetic nanoparticles showed the highest removal capacity of arsenic. In the 12 wt% magnetic nanoparticles experiment, almost 40% of the arsenic ion was removed in a single batch [19].

Moghaddam et al. further modified the magnetic Fe_3O_4 nanoparticles. They protected the nanoparticles against oxidation by providing a APTES layer. They used the modified nanoparticles and doped the polyurethane foam with different (3, 6, 8, 12, 21) weight percentages [8,20–22].

5.4.2.1 Method of Production

To prepare the PNC in the lab, 1 gm of Fe_3O_4 nanoparticle and 250 mL of ethanol were continuously stirred in the beaker for 10 min. Some acetic acid was added, and 4 mL of APTES was added to modify the nanoparticles. Now an inert atmosphere was created using N_2. Under the N_2 atmosphere and room temperature, the mixture was stirred for 5 h. An external magnet was used to separate the modified nanoparticles. After washing the nanoparticles with distilled water, the nanoparticles were dried under vacuum conditions at a temperature of 40°C. Then, these nanoparticles were distributed throughout the polyol matrix. The matrix was added with 0.1 mL of distilled water as a blowing agent. Finally, methylenediphenyl diisocyanate and the sample were mixed in a mold at the ratio of (polyol: isocyanate) 24:17 weight ratio. The mixture was kept at room temperature for 24 h. It finally formed the PNC foam [8,19,20].

Moghaddam et al. conducted batch experiments using varying concentrations of the nanocomposite. With 100 ppb of arsenic solution kept in 50 mL tubes, 1 gram of the nanocomposite foam was placed and then made to wait for 3–5 h Room temperature was maintained, and a pH of 6.5 was maintained. It was observed that the best results were shown by 12 wt% and 20 wt% nanocomposites. The 12 wt% showed 75.65% removal capacity, and the 20 wt% nanocomposites showed 92.81% removal capacity [8].

The experiment made it evident that the APTES layer enhanced the performance of the nanoparticles.

5.4.2.2 Adsorption Isotherm

Moghaddam et al. collected data from the batch experiment, and with these data, many conclusions were drawn about the adsorption isotherms and adsorption kinetics. The Langmuir and the Freundlich isotherms were plotted. It was observed that at a low concentration of the nanoparticles (8 wt%), the regression coefficient (R^2) value was 0.65. However, when the concentration of the nanoparticles was increased to (20 wt%), there was a significant rise in the R^2 value to 0.93. Thus, it shows that the nanoparticle's adsorption kinetics varies a lot when its concentration is varied. The Freundlich isotherm showed that a value of the Freundlich constant (n) > 1 showed favorable adsorption.

5.4.2.3 Adsorption Kinetics

The adsorption kinetics was derived by collecting the data from the batch experiment by varying the contact time. Then the data was plotted and checked whether the experimental results observed pseudo first-order or pseudo second-order kinetics.

The equation defines the pseudo first-order kinetics

$$\log(c_e - c_t) = \log\left(c_e - \frac{K_1}{2.303}\right)t \qquad (5.5)$$

In the equation, c_e (mg/g) is the maximum adsorption capacity.

c_t (mg/g) is the amount of solute adsorbed at a time t.
K_1 is the pseudo first-order constant.

With the R^2 value around 0.695, it was evident that the results did not follow pseudo first-order kinetics.

The equation for the pseudo second-order kinetics:

$$\frac{1}{c_t} = \left(\frac{1}{K_2 c_e^2}\right)t + \frac{1}{c_e} \qquad (5.6)$$

The R^2 value was pecked at 0.946, signifying that the system followed pseudo second-order kinetics [8].

5.4.3 Cd^{2+} Cadmium (II) Ion Removal

The average concentration of cadmium in the soil is nearly 0.1–0.5 parts per million (ppm), but the World Health Organization (WHO) has set the recommended limiting range at 0.003 mg/L for Cd (II) in drinking water [21]. Cadmium is very common among all toxic heavy metal ions. It also has a carcinogenic effect even at a very low interaction limit. Long-term exposure to cadmium has the worst impact on living organisms like osteomalacia, kidney, liver damage, and respiratory and cardiovascular problems. Therefore, to avoid cadmium poisoning, its removal from the water gets immense importance [23–26].

Various treatment methods have been lab-tested, tested, and commercialized for Cd (II) removal, with the major ones being adsorption [27]), precipitation, ion exchange [28], and electrodialysis [29]. Most of these techniques have their restrictions in incomplete extraction, production of secondary wastes, and very high operating costs. Various adsorbents captured attention to reduce operational cost due to their abundance and easy handling. There are a lot of Cd adsorbents that have been studied in recent years, such as types of zeolites and clay materials, activated carbon blocks, types of substituted polymers, metal oxides, oxide of graphene, metalorganic frameworks and two-dimensional carbon nanotubes, etc. the above-mentioned species were used due to having higher surface area, higher abundance and higher chemical and mechanical stability. PNC were also used for cadmium removal because reusability in the practical application was the added advantage [26,30].

In an experiment, M. Sayed and N. Burham [32] produced a new nanocomposite adsorbent by polymerizing in the presence of 5% by weight of organobentonite/iron oxide instead of toluene diisocyanate and polyol Polyurethane foam/organic bentonite/iron oxide. As a polymer matrix possessing a considerable surface area, it is also resistant to acids, alkalis, and organic solvents. The lowest cost and structural stability make it more suitable.

5.4.3.1 Preparation Technique

To a 100 mL liquid mixture containing $FeCl_3$ (2.0 mol/L) and $FeSO_4$ (1.0 mol/L), urea solution was added (2.0 mol/L) under continuous stirring at 500 rpm at a pH around 11–12 to prepare iron oxide in the initial step, which was then rinsed several times to make it reach a pH value of 7 and then oven-dried for 8 h at a temperature of 80°C for 8 h. It was then ground and stored in a desiccator after passing through a 140 μm sieve.

Then the simultaneous insinuation of both hexadecyltrimethylammonium bromide (HDTMAB) and IO resulted in the production of *Organobentonite-iron oxide*. At 70°C, 5 g of Na–B was added to 85 mL DDW with magnetic stirring until the product was evenly dispersed. The prepared solution was added to the bentonite suspension at that same temperature. It was followed up by mechanical stirring at 1,000 rpm for an hour. The product was vacuum-filtered and washed with DDW numerous times until the bromide ion was removed (checking bromide with 0.1 N AgNO3 solution). It was then oven-dried for 12 h at 60°C before being sent for grinding. Finally, it was sieved to 140 microns and kept in a desiccator.

Organobentonite/iron oxide was prepared using polyurethane foam in the final stage. First, 20 g polyol was mixed with 1.9 g OB/IO (5 wt%), followed by a few drops of stannous octoate with polycarbonate, and constant stirring was performed. The dispersion was then added with a solution comprising 13.5 g TDI, two drops of silicone oil, and 2 mL distilled water under continuous and vigorous stirring. The liquid polymer was placed into a mold as soon as the foaming began and held at ambient temperature for the rest of the day. The foam material was divided into 0.2 cm side cubes, and then it was rinsed overnight in distilled water. Subsequently, the excess inorganics were removed by rinsing with 0.1M HCl, followed by washing with acetone to separate the residual organic compounds. Finally, it was rinsed in distilled water and dried under normal conditions.

5.4.3.2 Adsorption Experiments

A series of adsorption experiments were performed where a 25 mL working mixture containing cadmium ion was introduced to a 250 mL polyethylene volumetric flask. The adsorption kinetics was evaluated using a 200-rpm orbital shaker by varying the uptake duration from 2 to 90 min. Furthermore, by increasing the starting metal concentration from 0.6 to 100 ppm, the adsorption isotherm was computed. An atomic absorption spectrophotometer was used to quantify the excess Cd (II) ion concentrations. The initial (C_o) and equilibrium ($C_{eq.}$) cadmium ion concentrations can be used to calculate the quantity of Cd (II) adsorbed.

The % removal of Cd (II) is obtained from the difference between C_o and C_{eq} using the following expression:

$$qe \text{ and } \%Cadmium \text{ } (II) \text{ } Removal = \qquad (5.7)$$

Here q_e is mg of Cd (II) adsorbed (mg/g), C_o and C_e are the initial and equilibrium concentration respectively, m describes the mass of the adsorbent in (g), and V is the volume of the Cd(II) solution in (L).

5.4.3.3 Regeneration of Adsorbent

The regeneration of polyurethane foam nanocomposite was measured by conducting reusability experiments and selecting an effective desorption reagent. So, 0.05 g nanocomposite was stirred at 200 rpm for an hour at neutral pH with a Cd (II) solution concentrated 100ppm. The washing was done using DDW on the samples, followed by drying. Following this, different eluting reagents like HCl, HNO_3, or H_2SO_4 were used to treat each of the samples prepared, all at the same volume of 25 mL with 0.1M concentration. The adsorbent had to be washed with DDW until it became acid-free and ready to be used for the next run. For proper evaluation, six consecutive adsorption and desorption cycles were observed repeatedly.

5.4.3.4 Adsorption Kinetics

It was confirmed by the kinetics study of the process that the adsorption of cadmium metal ions on the nanocomposite foam followed the chemisorption mechanism [29]. Scientists indicated that the liquid film diffusion step might be a part of the rate-determining step (RDS) and M. Sayed and N. Burham concluded that this cadmium ion adsorption process onto the nanocomposite was a three-step process [32]. Initially, diffusion through the external liquid film was followed by relatively sluggish pore diffusion, and cadmium ions were eventually adsorbed onto the active sites. Also, the results of the kinetic studies show that the removal of cadmium in nanocomposites follows a quasi-second-order kinetic model.

5.4.3.5 Adsorption Isotherm

M. Sayed and N. Burham [31,32] collected data from the batch experiment, and with these data, the adsorption behavior of the systems was drawn about the adsorption isotherms and adsorption kinetics.

$$q_e = \frac{Q_b \times C_e}{1 + bC_e} \quad \text{(for Langmuir isotherm)} \tag{5.8}$$

Here q_e, Q and b are equilibrium adsorption capacity (mg/g), maximum monolayer capacity (mg/g), and enthalpy of adsorption (L/mg), respectively.

$$q_e = K_F + C_e^{\frac{1}{n}} \quad \text{(for Freundlich isotherm)} \tag{5.9}$$

Here K_F is the unit capacity coefficient, and "n" is the Freundlich constant related to the system heterogeneity.

With the above-mentioned equations used to plot the isotherm graphs, the Langmuir isotherm was established to be the best fit model, with a regression coefficient ($R^2 = 0.993$) comparatively higher than that of Freundlich's ($R^2 = 0.96$) with the convergence between the estimated values of q_m and experimental readings. M. Sayed and N. Burham [31,32] also confirmed that the chemisorption mechanism favored Cd (II) adsorption using nanocomposite foam.

5.5 CONCLUSIVE REMARKS

It is quite evident that the presence of metal ions is a huge threat to the ecological surroundings and living organisms, as they are non-biodegradable. Although many methods like reverse osmosis, chemical precipitation, and electrochemical treatment may be used, they need to be cost-effective. Polymeric foams are extremely important as they can be tuned to have degradable properties and are environmentally friendly. However, the PNC foams show way better metal removal efficiency. Recent studies reflect that the inclusion of nanoparticles enhances the specific surface area of polymeric foam. Although normal polymeric foams find it challenging to separate metal ions from their aqueous solution due to their microscopic size, nanocomposite foams can efficiently remove metal ions due to their nanoscale size. The proposal of nanocomposite foams seems to be a promising approach to removing the metal contaminants present in the wastewater. However, the synthesis process of polymeric foams is quite time-consuming. Again, the commercial nanomaterials used for manufacturing nanocomposite foams for heavy metals removal on an industrial scale are rare to produce, and severe innovations are needed in this field to develop market-available nanocomposite foams. To commercialize nanocomposite foam, the synthesis and operating costs need to be optimized, and the requirement of green chemistry must be kept in mind for metal ions removal.

REFERENCES

1. Ashida, K. (2007). *Polyurethane and related foams: chemistry and technology.* Boca Raton, FL: CRC/Taylor & Francis.
2. Zhang, Y., Wu, B., Xu, H., Liu, H., Wang, M., He, Y. and Pan, B. (2016). Nanomaterials-enabled water and wastewater treatment. *NanoImpact* [online] 3–4, pp.22–39.
3. Klempner, D. and Kurt C.F. (2004). *Handbook of polymeric foams and foam technology.* Munich: Hanser Publishers.
4. Lee, S.-T., Park, C.B. and Ramesh, N.S. (2007). *Polymeric foams: science and technology.* Boca Raton, FL: CRC/Taylor & Francis.
5. Khemani, K.C., American Chemical Society. Division of Polymer Chemistry and American Chemical Society. Meeting (1997). *Polymeric foams: science and technology.* Washington, DC: American Chemical Society.
6. Throne, J.L. (1996). *Thermoplastic foams.* Hickley, OH: Sherwood Publishers.
7. Shutov, F.A. (1991). *Polymer foams: processing and production technology.* Lancaster, PA: Technomic Publishing Co.
8. Tamaddoni Moghaddam, S., Naimi-Jamal, M.R., Rohlwing, A., Hussein, F.B. and Abu-Zahra, N. (2019). High removal capacity of arsenic from drinking water using modified magnetic polyurethane foam nanocomposites. *Journal of Polymers and the Environment*, 27(7), pp.1497–1504.
9. Hong, H.-J., Lim, J.S., Hwang, J.Y., Kim, M., Jeong, H.S. and Park, M.S. (2018). Carboxymethlyated cellulose nanofibrils (CMCNFs) embedded in polyurethane foam as a modular adsorbent of heavy metal ions. *Carbohydrate Polymers* [online] 195, pp.136–142.
10. Kausar, A. (2018). Graphene nanoplatelet reinforced polyacrylonitrile/poly(vinylidene fluoride-co-hexafluoropropylene) nanocomposite foams: physical properties and ion detoxification. *Materials Research Innovations*, 24(1), pp.28–38.

11. Chavan, A.A., Li, H., Scarpellini, A., Marras, S., Manna, L., Athanassiou, A. and Fragouli, D. (2015). Elastomeric nanocomposite foams for the removal of heavy metal ions from water. *ACS Applied Materials & Interfaces*, 7(27), pp.14778–14784.

12. Dutta, B.K. (2009). Principles of mass transfer and separation processes. *The Canadian Journal of Chemical Engineering*, 87(5), pp.818–819.

13. Roy, S., Kargupta, K., Chakraborty, S. and Ganguly, S. (2008). Preparation of polyaniline nanofibers and nanoparticles via simultaneous doping and electro-deposition. *Materials Letters*, 62(16), pp.2535–2538.

14. Li, H., Zanella, M., Genovese, A., Povia, M., Falqui, A., Giannini, C. and Manna, L. (2011). Sequential cation exchange in nanocrystals: preservation of crystal phase and formation of metastable phases. *Nano Letters*, 11(11), pp.4964–4970.

15. Gao, H.-L., Lu, Y., Mao, L.-B., An, D., Xu, L., Gu, J.-T., Long, F. and Yu, S.-H. (2014). A shape-memory scaffold for macroscale assembly of functional nanoscale building blocks. *Materials Horizons Journal*, 1(1), pp.69–73.

16. Mark, J.E. ed., (2007). *Physical properties of polymers handbook*. New York, NY: Springer.

17. Puri, A. and Kumar, M. (2012). A review of permissible limits of drinking water. *Indian Journal of Occupational and Environmental Medicine*, 16(1), p.40.

18. Fryxell, G.E., Lin, Y., Fiskum, S., Birnbaum, J.C., Wu, H., Kemner, K. and Kelly, S. (2005). Actinide sequestration using self-assembled monolayers on mesoporous supports. *Environmental Science & Technology*, 39(5), pp.1324–1331.

19. Zhou, S., Wang, D., Sun, H., Chen, J., Wu, S. and Na, P. (2014). Synthesis, characterization, and adsorptive properties of magnetic cellulose nanocomposites for arsenic removal. *Water, Air, & Soil Pollution*, 225(5), p.8.

20. Zhu, J., Wei, S., Chen, M., Gu, H., Rapole, S.B., Pallavkar, S., Ho, T.C., Hopper, J. and Guo, Z. (2013). Magnetic nanocomposites for environmental remediation. *Advanced Powder Technology*, 24(2), pp.459–467.

21. Moyez, S.A. and Roy, S. (2017). Thermal engineering of lead-free nanostructured $CH_3NH_3SnCl_3$ perovskite material for thin-film solar cell. *Journal of Nanoparticle Research*, 20(1), p.88.

22. Hussein, F.B. and Abu-Zahra, N.H. (2016). Extended performance analysis of polyurethane–iron oxide nanocomposite for efficient removal of arsenic species from water. *Water Supply*, 17(3), pp.889–896.

23. Gunashekar, S. and Abu-Zahra, N. (2016). Structure–performance relationship of bulk functionalized polyurethane foams designed for lead ion removal from water. *Journal of Porous Materials*, 23(3), pp.801–810.

24. Zhu, J., Wei, S., Lee, I.Y., Park, S., Willis, J., Haldolaarachchige, N., Young, D.P., Luo, Z. and Guo, Z. (2012). Silica stabilized iron particles toward anti-corrosion magnetic polyurethane nanocomposites. *RSC Advances*, 2(3), pp.1136–1143.

25. Jane Wyatt, C., Fimbres, C., Romo, L., Méndez, R.O. and Grijalva, M. (1998). Incidence of heavy metal contamination in water supplies in Northern Mexico. *Environmental Research*, 76(2), pp.114–119.

26. Chen, Y.-G., Ye, W.-M., Yang, X.-M., Deng, F.-Y. and He, Y. (2010). Effect of contact time, pH, and ionic strength on Cd(II) adsorption from aqueous solution onto bentonite from Gaomiaozi, China. *Environmental Earth Sciences*, 64(2), pp.329–336.

27. Maitra, S., Sarkar, A., Maitra, T., Halder, S., Kargupta, K. and Roy, S. (2021). Solvothermal phase change induced morphology transformation in $CdS/CoFe_2O_4@$ Fe_2O_3 hierarchical nanosphere arrays as ternary heterojunction photoanodes for solar water splitting. *New Journal of Chemistry*, 45(28), pp.12721–12737.

28. Lin, X., Burns, R.C. and Lawrance, G.A. (2005). Heavy metals in wastewater: the effect of electrolyte composition on the precipitation of cadmium (II) using lime and magnesia. *Water, Air, and Soil Pollution*, 165(1–4), pp.131–152.

29. Kocaoba, S. (2007). Comparison of Amberlite IR 120 and dolomite's performances for removal of heavy metals. *Journal of Hazardous Materials*, 147(1–2), pp.488–496.
30. Oliveira, L.C.A., Rios, R.V.R.A., Fabris, J.D., Sapag, K., Garg, V.K. and Lago, R.M. (2003). Clay–iron oxide magnetic composites for the adsorption of contaminants in water. *Applied Clay Science*, 22(4), pp.169–177.
31. Sayed, M. and Burham, N. (2017). Removal of cadmium (II) from aqueous solution and natural water samples using polyurethane foam/organobentonite/iron oxide nano-composite adsorbent. *International Journal of Environmental Science and Technology*, 15(1), pp.105–118.
32. Burham, N. and Sayed, M. (2016). Adsorption behavior of Cd^{2+} and Zn^{2+} onto natural Egyptian Bentonitic clay. *Minerals*, 6(4), p.129.

6 Polymeric Nanocomposite Foams in Biomedical Engineering

Vipul Agarwal

CONTENTS

DOI: 10.1201/9781003218692-6

6.1 INTRODUCTION

Polymer nanocomposites continue to draw significant research interest due to the ease of manipulating the intrinsic properties of the polymer matrix with the inclusion of a small amount of filler material. The vast applicability of polymer nanocomposites in a wide range of fields has motivated researchers to transform them into functional three-dimensional (3D) foams. Polymer nanocomposite foams describe the low-density, lightweight 3D porous matrix loaded with fillers exhibiting summative properties of the polymer and filler. The accelerated growth in polymer nanocomposite foams has been fueled by the development of neat polymer foams and advancements in nanotechnology, both of which have independently penetrated our everyday life. Considering the wide variety of polymers fabricated into 3D foams and armory of different types of fillers including nanoparticles, the potential combinations for nanocomposite foams are only limited by imagination and technical advancements in terms of new fabrication methodologies.

The 3D polymer nanocomposite foams have found applications in different domains including sensing, water filtration, electromagnetic shielding and biomedical science and engineering (He et al., 2020, Panahi-Sarmad et al., 2020, Stucki et al., 2018, Singh et al., 2018, Wang et al., 2018). This chapter is focused on the application of 3D polymer nanocomposite foams for specific applications in the biomedical field. The detailed review of all different application areas is outside the scope of this chapter. In this chapter, the first nanocomposites and nanocomposites foams will be introduced, followed by a description of their fabrication strategies, different characterization techniques and application in biomedical engineering and finally, a conclusion and future outlook for the field is presented to motivate and drive future research in this field.

6.2 POLYMER NANOCOMPOSITES

Polymer nanocomposites are widely used in automotive, construction, electronics and aerospace industries, primarily due to their improved properties (mechanical, electrical, physical and surface) compared to neat polymer counterparts (Lee et al., 2005). Within polymer nanocomposites, there has been a growing inclination towards the use of nano-size fillers. Polymer nanocomposites comprising nanoparticulate fillers have been shown to exhibit superior properties without compromising the density and lightweight profile of the polymer matrix as compared to micron-size fillers. The nanoparticulate filler in the nanocomposite can have one, two or three dimensions on a nanometer scale. For example, platelets with one dimension on a nanometer scale (clay, graphene and derivatives, transition metal dichalcogenides and MXenes), nanotubes and nanofibers with two dimensions on a nanometer scale (carbon nanotubes (CNTs), carbon nitride tubes, carbon nanofibers, and polymer worms), spherical particles with all three dimensions in a nanometer scale (metal and polymer) (Agarwal and Chatterjee, 2018, Agarwal et al., 2019, Delafresnaye et al., 2017, Kausar, 2020, Lee et al., 2005, Li et al., 2019, Mallakpour and Rashidimoghadam, 2017, Mittal, 2014, Meng et al., 2020, Tran et al., 2021a, 2021b, Zabihi et al., 2018, Zare and Shabani, 2016). Over the years numerous synthetic and fabrication strategies have been developed to formulate polymer nanocomposites including polymer processing approaches (such as melt mixing and solution mixing), in situ polymerization (emulsion-based polymerization, melt polycondensation) (Fadil et al., 2022). The same fabrication strategies or their derivatives have been adopted to fabricate polymer nanocomposite foams, which will be elaborated on and discussed in this chapter.

For biomedical applications, a plethora of different fabrication strategies has been developed from simple approaches like drop casting, spin coating and layer-by-layer assembly to more sophisticated methods such as lithography, electrospinning and (bio)printing (Agarwal et al., 2017, Agarwal et al., 2021, Derakhshanfar et al., 2018, Eroglu et al., 2012, Ho et al., 2015, Kishan and Cosgriff-Hernandez, 2017, Li et al., 2018, Liang et al., 2007, Meka et al., 2019, Maslekar et al., 2020, Melchels et al., 2010, Zuev et al., 2020, Ziminska et al., 2019). The inclusion of nanofiller within the polymer matrix provides additional advantages over neat polymer counterparts such as surface adhesion sites (promote cell proliferation and motility), polarity (could induce differentiation of progenitor or stem cells), mechanical properties (recapitulating innate tissue environments), electrical properties (induce linage-specific differentiation in stem cells) and antimicrobial properties (Bužarovska et al., 2019, Huang et al., 2021, Li et al., 2018, Meka et al., 2019, Samantaray et al., 2020). Based on these innumerable advantages, biomedical research has been directed towards combining these advantages with high porosity in the form of nanocomposite foams.

6.3 POLYMER NANOCOMPOSITE FOAMS

The architecture of polymer nanocomposite foams akin to neat polymer foams comprises highly porous interiors made of gaseous voids surrounded by denser matrix. Foams with high porosity and interconnected pore cell network have been central to tissue engineering applications due to their ability to promote cell attachment and growth. Nanocomposite foams can be classified based on the size of pore cells – macrocellular (>100 µm), microcellular (1–100 µm), ultramicrocellular (0.1–1 µm) and nanocellular (0.1–100 nm) (Lee et al., 2005). In addition, nanocomposite foams can also be defined as open or closed-cell foams. In open-cell foams, cell walls are broken and intertwined and comprise mainly struts and ribs. In closed-cell foams, the compartments (cells) are more defined and isolated from each other, with cavities surrounded by complete cell walls. The open-cell foams are softer and more absorbent than closed-cell foams, which tend to be more rigid with insulating properties. By manipulating the cell structure and interplay between the two types (open and close cells), the final properties and microstructure of the nanocomposite foams can be tailored (Chen et al., 2013). In biomedical applications, open-cell foams are preferred due to their higher permeability compared to closed-cell foams. Higher permeability is a highly desirable feature for tissue regeneration and is considered necessary for nutrient transport and exchange of gases required to support cell ingrowth leading to the formation of functional tissue.

The inclusion of nanofillers within a polymer nanocomposite foam offers tailor ability towards their final properties. For example, nanofillers act as heterogeneous nucleation sites, where due to their small size (more numbers per unit volume) they provide a large number of nucleation sites. The increased number of nucleation sites commonly leads to smaller cell size, which in turn affects the properties of the foam. The addition of nanofillers can also introduce new functionalities such as electrical and thermal conductivity, while simultaneously enhancing mechanical strength. Nanofillers can also change the rheological behavior of the innate polymer by increasing the viscosity of the nanocomposite, which inhibits void coalescence and consequently decreases the average cell size when the nanocomposite is molded into a foam. In addition, nanofiller surface chemistry can be adjusted to control their arrangement and distribution (which can also be manipulated by the introduction of a surfactant) by improving polymer–filler interactions (Fadil et al., 2019, 2022), which can reduce the energy barrier for bubble nucleation in gas foaming methods allowing altercation in structure and density of the foam (Chen et al., 2013). The extensive list of factors that can influence the morphology of the nanocomposite foams has been previously reviewed in great detail (Chen et al., 2013).

6.3.1 WHY 3D NANOCOMPOSITE FOAMS FOR BIOMEDICAL APPLICATION?

Nanocomposite foams make it possible to mimic nano/microscale topography along with the mechanical, surface, chemical and biological properties of the extracellular matrix (ECM). Compared to neat polymer foams, the inclusion of a filler allows precise control and tailor ability in almost all the properties of the nanocomposite foam. For example, the incorporation of a filler increases the applied surface area and provides contact (adhesion) points between the cells and the matrix surface for better information transfer. In the case of nervous, cardiac and musculoskeletal systems, the inclusion of an electrically conductive filler is highly desirable to facilitate the smooth transmission of signals between, for example, neuronal cells.

Multiscale porosity is preferred in bio scaffolds due to the reason that different size pores serve different functions. Small, well-connected biological cell length-scale pores promote cell adhesion, proliferation and migration, microporosity with pore size between 1 and 50 μm is required for nutrient diffusion and metabolic waste transport (Salerno et al., 2011a), while large pores (minimum 30 μm and ideally >300 μm have been reported as beneficial for osseous tissue deposition (Owen et al., 2020).

6.3.1.1 Fabrication of Polymer Nanocomposite Foams

Nanocomposite foams can be fabricated by using different methods some of which have been described in this section. In the foam industry, the majority of large-scale commercial foam production is based on the use of blowing agents. These blowing agents introduce gas into the polymer matrix, and these gas bubbles when released introduce porosity in the foam. The gas phase can be either produced in situ during the polymer synthesis process (like in polyurethane (PU) foams and supercritical carbon dioxide method) or mechanically introduced (gases like nitrogen, argon and carbon dioxide).

In Situ Polyurethane Foaming Method

In this method, first polyols (like PEG and glycerol ethoxylate) are mixed with an isocyanate species, surfactant and filler (which can be modified to also act as an isocyanate species) (Xie et al., 2017) at a constant temperature. Next, the foam formation is initiated with the addition of a catalyst (such as KKAT-XK-627, dibutyltin dilaurate (DBDTL)) in water, and finally cured at a fixed temperature to obtain a robust foam. Traditionally, the isocyanate and polyol ratios are maintained at 1:1. In this method, catalysts should be carefully selected especially if the foams are designed for biomedical applications. For example, the DBDTL catalyst when used in PU foam has been shown to cause a significant reduction in cell viability compared to KKAT (Lundin et al., 2017).

Freeze Casting (Drying) Method (also known as Ice Templating)

In this method, polymer soluble in a solvent is supplemented with a filler and mixed to obtain a uniform dispersion (solution mixing, *in situ* polymerization). The dispersion is then frozen either cryogenically (dipping in liquid nitrogen) or overnight in a freezer. The frozen dispersion is then lyophilized using a freeze dryer to sublimate and remove the solvent to obtain a dried nanocomposite foam (Deville, 2017, Scotti and Dunand, 2018, Yang et al., 2020). During the freezing step, filler and polymer (substrate) particles are expelled from the growing ice crystals and forced into the interstitial regions between the ice crystals, which controls the final architecture of the foam (Wang et al., 2017). The expulsion of the filler and polymer particles is highly complex and is dependent on interactions between the filler and polymer particles, their interactions with the solvent and ice, and interfacial interactions between the different phases (Wang et al., 2017). This is a relatively simple method, which can be adapted to a range of polymers and fillers, where the pore structure and internal morphology can be tailored by controlling the freezing direction, rate and temperature, interactions between the filler and polymer particles and their weight fractions. However, colloidal stability of the two components is a prerequisite for this approach, and the requirement of aqueous solvents in the freeze dryer limits the type of polymers which can be explored (Deville, 2017). The advantages of this method include polymer/filler dispersion can be directly used without requiring any crosslinking or prior processing, do not require any post-processing (rinsing), and homogeneity of the foam can be achieved by maintaining homogeneous freezing.

Porogen Leaching

First, polymer along with the filler is dispersed in an organic (hydrophobic) solvent (such as dimethylformamide or chloroform), which is then added to a bed of salt (sodium chloride or sodium bicarbonate) template to uniformly impregnate the template (centrifugation can be used to aid impregnation). The impregnated polymer dispersion is dried at room temperature and the salt template is removed by washing under water to obtain a porous foam with nearly uniform pore size with unimodal distribution. The porosity and pore size can be controlled by varying the amount and size of the salt crystal, respectively used in the template. The advantage of this method includes the use of small amounts of polymer minimizing wastage. However, this method can cause limited pore interconnectivity due to the lack of contact between the porogen crystals to have a continuous porosity (Owen et al., 2020). Recently, to achieve bimodal pore distribution, this approach has been further advanced by incorporating two porogens (sugar and salt) in the polymer/nanofiller dispersion (Jafari et al., 2018Padash et al., 2021, Shahrousvand et al., 2017). In this case, salt is used to form the template, while sugar is mixed with the polymer/filler nanocomposite matrix (in organic solvent) before adding it to the salt template. Washing of the salt template also removes sugar to form bimodal porous foam (with pore structure from tens of nanometers to hundreds of microns). Due to high reproducibility of the obtained foams, this method has been widely explored for biomedical applications. The biggest limitations of this method include (i) being restricted to only hydrophobic polymers and (ii) being difficult to control the leaching of filler during washing and biological studies.

Thermally Induced Phase Separation (TIPS)

This method is based on thermally induced phase separation (TIPS). Polymer (dissolved in a good solvent and a non-solvent) and a filler are mixed and maintained at a fixed temperature (to obtain homogenous dispersion), followed by rapid lowering of temperature close to the cloud point of the polymer within its metastable region for 10–30 min. Next, the dispersion is rapidly quenched by immersion in an ethanol bath at $-20°C$ to freeze the phase-separated structure to obtain the foam (Carfì Pavia et al., 2018). This method leads to small pore sizes, while the addition of organic solvents inhibits the use of bioactive molecules.

Foam Injection Molding and Mold Opening

The prerequisite to this method is the melt mixing to incorporate a filler in the polymer matrix and extrusion steps. In this approach, extruded pellets are fed into an injection molding machine, at a high temperature (270/320°C), followed by injection of a fixed amount of blowing agent (N_2 gas), which is mixed with the polymer melt to form a uniform polymer/gas mixture and maintained at high temperature (200°C). The mixture is then injected at a high flow rate into a mold (to avoid gate-nucleated cells) at high mold cavity pressure; the mold temperature is subsequently allowed to cool to 50°C for a very short dwell time (45 s). Finally, the mold is opened to obtain a foam with high void fractions (Lee et al., 2020). This method is primarily applicable to high glass transition temperature polymers and may provide only limited control over the arrangement and distribution of the filler within the foam polymer matrix.

Supercritical Carbon dioxide (Sc-CO₂) Foaming

In this approach, polymer and filler are melt mixed in an extruder, followed by compression molding to form a disk. The molded composite disk is soaked to complete saturation in Sc-CO$_2$ in a high-pressure autoclave under heating. After Sc-CO$_2$ saturation, the temperature in the autoclave is cooled to polymer foaming temperature, which is maintained for a short time, and pressure is rapidly released to induce foam formation and brought to ambient temperature. The pressure drop rate controls cell nucleation and therefore, needs to be carefully controlled to tailor foam porosity and internal morphology. Similar to the foam injection molding method, this approach is limited to the type of polymers that can be used. This approach is suitable for amorphous polymers like polystyrene due to the ease of diffusion of blowing agent in amorphous glassy phase than in crystalline domains present in crystalline and semi-crystalline polymers such as poly-ε-caprolactone (PCL) (Salerno et al., 2011a).

Further, it is difficult to control the pore connectivity and pore sizes, which rely on solubilization pressure (gas concentration), foaming temperature and pressure drop rate (Salerno et al., 2011a, Qiu Li Loh and Choong, 2013). It has been reported that cell size (diameter) is dependent on the foaming temperature, where an increase in foaming temperature caused an increase in cell diameter (Moghadas et al., 2020). It was reasoned that an increase in temperature reduced the viscosity of the polymer/ filler melt, which promoted the movement of polymer chains in the softened polymer matrix. Softening of the polymer matrix induces less resistance to polymer expansion during foaming, thus increasing the growth of nuclei (nucleation point). Further, increasing the temperature reduces Sc-CO_2 solubility within the polymer matrix, which consequently reduces the nucleation rate leading to a reduction in the number of cells while increasing the diameter of the forming cells (Moghadas et al., 2020).

Batch Foaming

Like the Sc-CO_2 method, in this approach polymer/filler mixture is first saturated with a foaming agent (which could also be an inert gas) under certain temperatures and pressures. The saturation temperature controls the cell nucleation rate. The temperature is maintained above the Tg of the polymer and the pressure is rapidly released to force supersaturation, cell nucleation and growth to obtain a 3D foam. The cell density of the foam is dependent on the saturation temperature, pressure, and pressure drop rate (Lee et al., 2005).

High Internal Phase Emulsion (HIPE)

This method is different from all other fabrication methods. In this *in situ* foam fabrication approach, two immiscible liquids (organic and aqueous) are mixed in the presence of a surfactant, which also functions as an emulsifier. The continuous or external phase comprises of monomer, surfactant and an initiator, which is supplemented with water as an internal phase to form a stable emulsion. Polymerization of this emulsion result in foam formation, the porosity of which is dependent on the amount of water used and water droplet size (Silverstein and Cameron, 2010). Poly high internal phase emulsion (HIPE) is one of the most widely used industrial approaches to fabricating commercial polymer foams. One of the biggest limitations of polyHIPE is the difficulty in incorporating bioactive molecules and fillers (have to be homogenized in monomer solution), challenging their wider applicability in biomedical research.

Mixed Methods

there has been a growing interest in developing modified approaches by combining different fabrication methods, in particular, to introduce multimodal porosity in nanocomposite foams. For example, Salerno et al. presented a modified ScCO_2 approach to obtain bimodal pore distribution in PCL/hydroxyapatite (PCL/HA) foams (Salerno et al., 2011a). In their study, PCL was melt mixed with HA and compression molded to obtain 200 µm thick films, which were subjected to ScCO_2 equilibration at 20 MPa and 37°C for 1.5 h, followed by slow release of pressure to an intermediate pressure between 6 and 10 MPa (non-isothermal) to promote nucleation and growth of larger pores. The temperature of the system was then increased again to 37°C for ~5 min, after which the pressure was rapidly released within 1–2 s to promote nanopore formation leading to bimodal pore size distribution (Salerno et al., 2011a). In an alternate strategy gas (CO_2 – N_2) foaming method was combined with porogen leaching to obtain multimodal pore size distribution. In this combinatorial approach, PCL/HA was supplemented with NaCl salt during the melt mixing step. The obtained composite (containing NaCl) was compression molded and processed first through the gas foaming method (single decompression method) to form large pores. The embedded salt within the foam was extensively washed with water to induce small pore formation (Salerno et al., 2011b). It was reported that the extent and distribution of microporosity of the foams were dependent on the amount of CO_2 and salt used, where microporosity increased with increasing CO_2 amount but reduced with increasing salt concentration. More recently, polyHIPE was combined with porogen leaching to fabricate multimodal porous foams (Owen et al., 2020). In this study, instead of traditional salt or sugar, authors used alginate beads of uniform size (~700 μm) as porogen. They first formed the emulsion by mixing the continuous phase comprising monomer 2-ethylhexyl acrylate with a crosslinker trimethylolpropane triacrylate, a surfactant Hypermer B246-SO-(MV) and a photoinitiator (diphenyl (2,4,6-trimethylbenzoyl) phosphine oxide/2-hydroxy-2-methylpropiophenone) with the internal phase made up of water. The emulsion was then supplemented with alginate beads to form a homogeneous dispersion. Polymerization and foam formation of the continuous phase were conducted by exposing the dispersion to UV light (100 W). The alginate was leached by soaking the foam in sodium citrate aqueous solution to introduce microporosity. The obtained foam comprises pore size from 1 to ~780 μm (Owen et al., 2020).

6.3.2.2 Characterization of Polymer Nanocomposite Foams

The polymer nanocomposite foams are characterized using a range of techniques for specific properties. Considering the well-established correlation between the foam's intrinsic properties and cell behavior, thorough characterization of the nanocomposite foams is warranted to avoid surprises post-implantation.

Morphology

The morphology of the foams can be imaged using scanning electron microscopy (SEM) and in cases where fillers have high electron density (e.g. metal-based fillers), micro-computed tomography (μCT) can be employed. For SEM imaging, samples must be coated with electrically conducting materials such as platinum, gold or carbon before imaging. In addition to surface morphology, SEM can also provide an estimation of the cross-sectional area, interconnectivity, and wall thickness. However, this requires either sectioning or cryo-fracturing of the sample to avoid compression-mediated damage to the foam before imaging.

μCT is an X-ray-based technique that can provide 3D morphology of intact foams. During imaging, thin 2D section images are taken, which when combined reveal the 3D morphology. The advantages of this method include the ability to image intricate details at high resolutions, non-invasive, does not require any coating or sectioning before imaging, no toxic chemical required for sample preparation and imaging, and it is non-destructive. However, μCT requires a long acquisition time and data storage space. Further, for this method to work, the filler is required to have a considerably different scattering length density (SLD) than the polymer, otherwise, the two components would be indistinguishable. To overcome this challenge, neutron scattering approaches (neutron tomography) can be employed, which although also require a difference in SLD between the polymer and the filler, but they are more tolerable and can provide meaningful data even in cases with relatively closer SLD values. The tomographic techniques allow imaging of large areas of the intact foam providing information on the exact arrangement of the filler within the polymer matrix, without requiring any coating. μCT and neutron scattering techniques also provide additional information on pore architecture, size and distribution.

Pore Size and Porosity

To determine the porosity and pore size of a nanocomposite foam, a range of techniques can be employed including Brunauer-Emmett-Teller (BET), mercury porosimetry, nuclear magnetic resonance (NMR), gravimetry, and liquid displacement method.

In the gravimetry method, the total porosity (P) of the nanocomposite foams can be calculated using the bulk and true density of the material by employing the expression

$$P = 1 - \frac{d_f}{d_s}$$

where P represents total porosity, d_f is the fabricated density of the foam and d_s represents the density of the solid polymer. The density of the foam can be calculated from the mass and volume (dimensions) (Parai and Bandyopadhyay-Ghosh, 2019, Wang et al., 2019). Despite the simplicity of this method, it provides a rough estimate of the actual porosity due to possible errors which can incur when determining the actual volume of the foam. This method is useful for soft foams, which cannot withstand high pressure used in other measurement techniques.

Mercury porosimetry can be used to obtain the total pore volume fraction, average pore diameter, and pore size distribution of nanocomposite foams (Qiu Li Loh and Choong, 2013). In this method, foams are impregnated with mercury under increasing pressures (up to a maximum of 414 MPa) to ensure complete penetration of mercury into the pores. Higher pressures are required to fill small pores due to higher surface tension caused by the greater curvature of the surface meniscus. Although this method yields relatively more reliable results, it has a number of disadvantages including toxicity and cost of mercury, and applicability to only hard foams (soft foams with mechanical strength in tens of kPa collapse under high pressure required during measurements).

Another approach that can be used is the liquid displacement method where a non-solvent for the polymer is employed (e.g. ethanol). In this method, foam is immersed using a known volume of a non-solvent in a measuring cylinder. To ensure complete penetration, a series of evacuation–repressurization cycles has to be performed. Total porosity can be calculated from the difference between the initial volume of the solvent and the volume left after the liquid-impregnated foam is removed divided by the difference between the total volume of the foam and the initial solvent, and volume left after the liquid-impregnated foam is removed (Qiu Li Loh and Choong, 2013).

NMR can also be used to determine total porosity and pore size distribution. The use of NMR for pore size determination has not been widely studied. In the NMR method, the foam sample is saturated with deuterated water (D_2O) under high pressure to penetrate small pores. The D_2O-saturated foam is subjected to an electromagnetic field and relaxation time is used to calculate porosity. However, potential interactions between the filler (hydrophilic) and D_2O can interfere with the results. In addition, it is important that the foam does not shrink in D_2O.

BET is another method that can be employed to measure the pore size and total surface area of the foam. The BET method utilizes nitrogen adsorption isotherm measurements at 77 K to calculate total surface area and pore size. Typically, a foam sample is exposed to nitrogen gas under increasing pressure. It is assumed that nitrogen gas can access the entire foam surface and has minimal interaction with both the polymer and the filler. The surface area measurements utilize Langmuir adsorption theory based on the assumption that gas molecules adsorb on the foam surface in infinite layers with no interlayer interactions (Palchoudhury et al., 2015). However, BET cannot be used for foams with micron size pore structures, while mercury porosimetry and NMR cannot be used for soft foams (mechanical strength in tens of kPa).

Mechanical Strength

For mechanical properties, dynamic mechanical analysis (DMA) is traditionally used, where the foam sample of specific dimensions and shape is compressed at either a fixed or variable rate to a fixed (strain) point to evaluate the mechanical behavior of the foam. Young's modulus is calculated from the initial slope from the elastic region of the stress-strain curve, while the compressive strength is the maximum stress a foam can bear before failure or at the highest strain.

Electrical Properties–

In the case of electrically conductive fillers, linear sweep voltammetry (LSV), cyclic voltammetry (CV) and electrochemical impedance spectroscopy (EIS) can be employed, where the electrical contacts to the foam can be made by using metal (gold, silver) coated plates as the two electrodes. The dried nanocomposite foam is sandwiched between the two electrodes functioning as an anode and a cathode, which are connected to a potentiostat. It is important to maintain a constant temperature and humidity while taking the measurements as electrical conductivity directly correlates with ambient humidity (i.e. conductivity increases with increasing humidity). This correlation has been attributed to the ease of electrical transport facilitated by water molecules. The biggest error observed in this approach is the lack of proper contact made between the foam and the two electrode plates.

6.4 IMPORTANCE OF FOAM PROPERTIES IN BIOMEDICAL APPLICATION

6.4.1 MORPHOLOGY

Foam architecture should mimic the ECM to recapitulate the native tissue environment necessary to support cell adhesion, proliferation, migration and differentiation. The nanofillers are aimed at providing nano/micro-scale surface topographies to regulate and control (cultured and native) cell behavior (Wu et al., 2014). The foam should also possess a similar macrostructure as the native tissue to support the growth of cells and new tissues after implantation. Furthermore, the morphology should not change upon implantation or long-term culturing of cells, and any potential change in morphology should be considered when designing a foam-based implant, especially for applications requiring long-term implantation.

6.4.2 PORE SIZE AND POROSITY

Porosity and pore size in the foam are crucial features required to facilitate nutrient transport and oxygen exchange, both of which are necessary for cell ingrowth. Pore size has been shown to incur cell type-specific response in terms of promoting or excluding cell ingrowth within a foam (Zeltinger et al., 2001). For example, scaffolds fabricated using the salt leaching method with a constant void fraction of 90% and different pore sizes (90 μm, >107 μm) were explored for cell ingrowth using different canine cells (microvascular epithelial cells, vascular smooth muscle cells and dermal fibroblasts). It was observed that microvascular epithelial cells were highly selective to the pore size and formed thin endothelial linings only in 90 μm pore-size foams. Vascular smooth muscle cells, on the other hand, formed uniform tissue specifically in >107 μm pore size foam, whereas dermal fibroblasts exhibited no pore size-dependent selective response in any of the foams studied (Zeltinger et al., 2001).

6.4.3 MECHANICAL PROPERTIES

Mechanical properties of nanocomposite foams should be carefully considered and precisely controlled. Mechanical strength can significantly influence the integration of the implanted nanocomposite foam and native surrounding tissue at the implant site, and thus manipulate the behavior of innate cells. For example, natural bone exhibits super-elastic mechanical properties with Young's modulus of 1–27 GPa, therefore, nanocomposite foam should mimic the strength, stiffness and mechanical behavior to avoid post-implantation stress shielding effects, which can result in implant failure (Wu et al., 2014).

6.5 BIOMEDICAL APPLICATION OF POLYMERIC NANOCOMPOSITE FOAMS

The research in synthetic biomaterials and tissue engineering has been focused on biodegradable polyester such as poly(lactide-*co*-glycolide) (PLGA), poly(L-lactide)

(PLLA), and PCL (Armentano et al., 2010, Agarwal et al., 2017, Boccaccini et al., 2010, Meka et al., 2018, 2019, Pina et al., 2015). This interest has been fueled by the intrinsic properties of these polyesters including (i) the ease of fabrication of these polyesters into porous scaffolds, (ii) good foamability, (iii) biodegradation by innate esterase enzymes, and (iv) production of non-toxic degradation products in biological studies (Liao et al., 2012). However, biodegradable neat polymers lack the mechanical strength demanded from a functional implanted scaffold. To this end, filers are commonly incorporated in neat polymers to not only improve but recapitulate the mechanical properties of the natural tissue. A similar combination of organic/inorganic composites is commonplace in natural tissues. For example, natural bone is a composite of organic collagen and inorganic apatites. Therefore, composites of biodegradable polymer and inorganic bone-forming filler (HA, bioactive glass) remain central to bio scaffolds developed for bone regeneration application.

6.5.1 BIOCOMPATIBLE NANOCOMPOSITE FOAMS WITH INORGANIC FILLERS

The Sc-CO_2 foaming method has been proven to be a promising method for the preparation of polymer/inorganic filler nanocomposite foams with tunable intrinsic properties to improve cell–matrix interactions. Salerno et al. prepared PCL/HA nanocomposite foams with bimodal and uniform pore size distributions by utilizing two-step depressurization in the Sc-CO_2 foaming method (Salerno et al., 2011a). The intermediate pressure had the most profound effect on pore size and their distribution, where intermediate pressure above 8 MPa led to bimodal pore size distributions while at 8 MP a resulted in highly interconnected unimodal pores. SEM study on hMSCs cultured on the bimodal porous foam revealed significant cell adhesion to walls of the nanocomposite foam forming a cell-to-cell communication network within 24 h after seeding (Salerno et al., 2011a). Subsequently, the same authors developed an alternate foam fabrication approach by combining salt porogen leaching with the gas foaming method (Salerno et al., 2011b). In this new study, sodium chloride was added to the PCL/HA mixture before the melt mixing step, and the temperature was controlled to preserve the shape and size of salt crystals. The obtained matrix was compression molded and solubilized in the mixture of CO_2–N_2 as a foaming agent at a fixed saturation pressure, which induced form formation with the release of pressure. The obtained foams were subsequently soaked in water to leach out salt and introduce multiscale porosity. The final foams exhibit highly interconnected fibrillar morphology with well-dispersed HA. The multiscale porous foam enhanced cell–matrix interactions with pre-osteoblasts MG63 cells initially adhering to the macropores (created by salt) at day 1 resulting in the formation of cell bridges between opposite pore walls. By day 14, a dense sheet of MG63 cells was observed covering the supporting scaffold (skeleton) architecture. In addition to promoting cell adhesion, porous scaffolds also enhanced the osteogenic differentiation of MG63 cells as determined with significantly high alkaline phosphatase (ALP) activity and calcium mineralization, both of which are markers for osteogenesis (Salerno et al., 2011b).

In another study, Pavia et al. fabricated biodegradable PLLA/HA nanocomposite foams using the TIPS method. Compared to the pure PLLA foam, the compressive strength of the PLLA/HA foam increased by more than double with the loading of only 30 wt% of HA (Carfì Pavia et al., 2018). However, a further increase in the HA content compromised the compressive strength due to the aggregation of HA particles within the matrix. A biocompatibility study revealed comparable cell viability in pre-osteoblastic MC3T3-E1 cells cultured on PLLA/HA and neat PLLA foams. ALP activity, which was used as an early osteogenic marker, was significantly upregulated in PLLA/HA foams compared to neat PLLA foams due to the osteoinductive effect of HA (Carfì Pavia et al., 2018).

Xie and co-workers developed a shape-memory PU/HA nanocomposite foam for bone regeneration application to develop a pre-cast foam that when implanted can take the shape of the injury site (Xie et al., 2017). They modified HA with isocyanate groups to function both as an isocyanate species (required to synthesize PU) and an inorganic crosslinker during foam formation to improve mechanical properties. The pore size, overall porosity and compressive strength increased with an increasing amount of isocyanate-modified HA with no change in pore interconnectivity. The increase in the nanocomposite compressive strength reached significance at 7% loading with a maximum value of ~14 MPa compared to neat PU (~9 MPa), which was attributed to isocyanate-modified HA crosslinking. A biocompatibility study conducted on murine MC3T3-E1 cells using foam-conditioned media revealed cell viability of around 95% in both nanocomposite and neat PU foams. The nanocomposite PU/isocyanate modified HA foams exhibiting body temperature shape memory behavior, which in conjunction with their mechanical properties being similar to the human trabecular bone (4–12 MPa) make these foams optimal for bone implant application (Xie et al., 2017). The ability of the shape-memory PU foam to take the shape of the injury site offers this approach a competitive edge over other polymers. However, more biological studies are required to determine the long-term efficacy and cytocompatibility of PU-based foams before these can be considered for clinical translation. In addition, the lack of biodegradability of PU foams also raises concerns about their potential translation into clinics.

While there are obvious advantages to using biodegradable polyesters, the inability to control the degradation rate in vivo persists to be one of the major drawbacks limiting their clinical translation. In an alternate approach, thiol-acrylate-based copolymers fabricated using the Michael addition reaction have been explored as biomaterial replacement (Cong Chen, 2015, Garber et al., 2013). For example, Garber et al. synthesized pentaerythritol triacrylate-co-trimethylolpropane copolymer (PETA/TMPTMP) using an amine-catalyzed Michael addition reaction (Garber et al., 2013). The PETA/TMPTMP/HA nanocomposite foam was prepared by mixing the synthesized copolymer with HA (20 wt%) and subjecting this mixture to compressed nitrous oxide (N_2O, used as a gaseous porogen). Copolymer foam without HA was fabricated under the same conditions and used as a control. Inclusion of HA caused no noticeable change in the pore size distribution between the nanocomposite and neat polymer foams. However, compared to the neat polymer foam, nanocomposite foam exhibited over five times higher compressive strength (~0.72 MPa). Biocompatibility study in human adipose-derived mesenchymal pluripotent stem cells (hASC) exhibited relatively higher metabolic activity in neat PETA/TMPTMP foam compared to the nanocomposite PETA/TMPTMP/HA foam and the PCL foam control (prepared using TIPS from 1,4-dioxane followed by lyophilization). Additionally, metabolic activity on all foam samples was significantly lower than TCPS. Total DNA activity (used as a proxy for cell proliferation and viability) was significantly higher in PETA/TMPTMP/HA and PCL foams (which were similar) than neat PETA/TMPTMP foam indicating that the inclusion of HA-induced potential osteogenesis in hASCs (Garber et al., 2013). It is widely accepted that differentiation and proliferation are competitive pathways and cells stop proliferating before undergoing differentiation. Chen et al. expanded this approach by fabricating PETA/TMPTMP/HA foams with different HA concentrations (0, 15, 20, 25, 30 wt% relative to the polymer) using the N_2O foaming method and compared their intrinsic properties against PCL/HA (20 wt%) control in an in vitro and in vivo bone defect repair model (Cong Chen, 2015). They observed the highest porosity of ~72% at 15 wt% HA loading and noticed a reduction in pore sizes with increasing concentration of HA resulting in foam with closed cell morphology at 30 wt% HA loading. It was suggested that increasing the HA content would cause an increase in the nanocomposite viscosity. Such an increase in the viscosity would reduce N_2O expansion and mobility during foaming resulting in smaller pore sizes. Contrary to the pore size, the compressive strength increased with increasing HA content, potentially due to the change in foam morphology from a porous to a compact structure. In vitro studies in hASCs revealed a reduction in cell metabolic activity and viability but an increase in their osteogenic differentiation with the inclusion of HA in both PETA/TMPTMP and PCL foams compared to neat polymer foams in osteogenic media. In vivo rat posterolateral lumbar spinal fusion study revealed significant bone defect repair response in PETA/TMPTMP/HA foam-treated animals through higher densification and endochondral ossification compared to neat PETA foam-treated animals 6 weeks after surgery (Cong Chen, 2015). This study highlights the clinical applicability of the gas foaming approach, where foams can be formed in situ at the injury site in the animal circumventing potential damage to preformed foams during implantation.

In another study, naturally occurring polysaccharide carrageenan was grafted with acrylic acid to form a biocompatible polymer matrix, which was mixed with nano-HA (n-HA) and graphene oxide (GO) and molded into a foam using the freeze casting method (Aslam Khan et al., 2020). The authors reported an inverse correlation between porosity and compressive strength with increasing concentration of GO in the nanocomposite foams. Inclusion of GO also promoted biodegradation of the foams due to increased hydrophilicity confirmed with increased swelling and water retention. In vitro study revealed spread cell morphology and cell viability of ~90% over 72 h in MC3T3-E1 cells in all conditions (foams with different GO loading). In addition to improved biodegradation, GO within the nanocomposite foam also provide anchoring sites for cells to adhere and take their characteristic spread morphology (Aslam Khan et al., 2020).

Other carbon derivatives such as carbon fibers (CFs) and CNTs have also been incorporated to function as reinforcers and to simultaneously improve the hydrophilicity of nanocomposite foams. Uddin et al. carried out covalent bonding of polyetheretherketone (PEEK) with CFs and CNT before blending in HA (Uddin et al., 2019). The blended PEEK/CNT (or CF)/HA foams were prepared by the salt porogen leaching method. The reinforcing effect of both CNT and CF was only observed at 0.5

wt% loadings. They reported a reduction in compressive modulus with an increase in CNT or CF loading beyond 0.5 wt%. Like other fillers, CNT and CF tend to agglomerate at high concentrations causing micro-phase separation within the polymer matrix leading to a reduction in mechanical strength at higher loadings. The authors also observed a change in the texture of the foams with the inclusion of carbon fillers from spongy in neat PEEK to relatively harder and less agile in nanocomposite foams. A biocompatibility study revealed better cell viability in PEEK/HA/CNT foam compared to PEEK/HA foam (Uddin et al., 2019). Despite promising results, bio-investigation of these foams was very premature and require further investigation to determine their true potential for bone regeneration application. Wang et al. fabricated shape memory PU/CNT foam using the sugar porogen leaching method as an implant to treat intracranial aneurysms (Wang et al., 2019). CNT was included to allow X-ray guided surgical implantation of the foam in endovascular operations at the injury site and induce shape recovery when stimulated with electrical stimuli via a resistive-heating mechanism. The PU/CNT foam exhibited almost complete shape recovery from a compressive strain of >50% within 2 min when subjected to 0.2 A applied electrical current (Wang et al., 2019). This report reinforces the potential of shape memory PU nanocomposite foams for implantations. However, no biological testing was performed to elucidate the biocompatibility of the developed nanocomposite foams and their utility to treat intracranial aneurysms.

Other examples of inorganic fillers include kaolin, nanoceria, TiO_2, calcium phosphate crystals and halloysite nanotubes (Eryildiz and Altan, 2020, Karadas et al., 2014, Karakoti et al., 2010, Lundin et al., 2017, Manoukian et al., 2019, Pelaseyed et al., 2020, Yang S, 2014). In one study, PU/kaolin nanocomposite foam was developed as a wound dressing material for hemolytic application (Lundin et al., 2017). Kaolin was added during the PU synthesis process to obtain nanocomposite foams. Kaolin had dose-dependent effect on the mechanical properties of the foam by increasing the tensile strength of the nanocomposite foams with increasing kaolin loading. The PU/kaolin nanocomposite foams exhibited high biocompatibility with cell viability between 85 and 90% in HeLa cells, which was similar to neat PU foam control. A hemolysis study revealed significant effect of kaolin in nanocomposite foams on blood clotting index and clotting time. PU/kaolin foams caused a significant reduction in clotting time compared to untreated natural blood and neat PU foam controls, making the nanocomposite foams a promising candidate for wound dressing material (Lundin et al., 2017). In another study, polylactic acid (PLA)/halloysite nanotube nanocomposite foams were fabricated using the foam injection molding method (Eryildiz and Altan, 2020). Inclusion of halloysite nanotubes increased the average pore density and tensile strength of the nanocomposite foams with increasing halloysite loading reaching the maximum values at 3 wt% (halloysite). However, both pore density and tensile strength reduced with a further increase in the halloysite loading (to 5 wt%) due to nanotube agglomeration. A biocompatibility study revealed relatively higher cell viability in L929 mouse fibroblasts at high halloysite loading (3 and 5 wt%) compared to the neat PLA foam. However, cell viability was a minimum of three times lower in nanocomposite foams than in TCPS control, the reason behind such low cell viability response in foams was not discussed. This low cell viability raises concern regarding the applicability of these PLA/halloysite foams and therefore requires further research. Recently, Manoukian et al. fabricated chitosan/halloysite nanotube nanocomposite foam-like conduit loaded with 4-aminopyridine with interconnected, longitudinally aligned pores using the unidirectional freeze casting method for peripheral nerve regeneration application (Manoukian et al., 2019). A biocompatibility study revealed significant proliferation of human Schwann cells on the nanocomposite foam over the period of 3 weeks while a confocal study showed full cell penetration with uniform distribution throughout the nanocomposite foam after 2 weeks in culture. In vivo study of critical-sized sciatic nerve defect, Wistar rats confirmed the saturability and strength of the nanocomposite foam to withstand ambulatory forces (including surgery) over 4 weeks of their implantation. Histological analysis of the excised foams revealed infiltration and orientation of the Schwann cells through the nanocomposite foam lumen and into the foam matrix (Manoukian et al., 2019). There continues to be a drive to control the directionality of neuronal cells by providing directional contact guidance cues to create long-distance neuronal connections damaged by trauma and disease to central and peripheral nervous systems (Ho et al., 2015, Li et al., 2018). The long-distance neuronal connections are required to restore function lost due to injury or disease to the nervous system. Therefore, the results from (Manoukian et al., 2019) provide a promising 3D implantable scaffold for nerve regeneration application.

Yang and co-workers prepared poly(3-hydroxybutyrate-co-3-hydroxyhexanoate) (PHBHHx)/siliceous mesostructured cellular (SMC) nanocomposite foams loaded with recombinant human bone morphogenetic protein 2 (rhBMP2) using the salt porogen leaching method and evaluated their efficacy both in vitro and in vivo in bone regeneration application (Yang S, 2014). SMC was employed as a replacement for mesoporous bioactive glasses, which have been widely accepted as effective osteoinductive fillers (Boccaccini et al., 2010, Meka et al., 2018, 2019), and size similarity with protein growth factors BMP2. They reported that the inclusion of SMC significantly reduced the release of rhBMP2 from the nanocomposite foam compared to neat PHBHHx foam due to hydrogen bonding interaction between SiOH of SMC and amino/hydroxyl groups in rhBMP2. A biocompatibility study revealed an increase in hMSCs proliferation with increasing SMC content. They also observed higher ALP activity in the PHBHHx/SMC foam compared to the neat PHBHHx foam with no change with increasing SMC content in the nanocomposite foams. In vivo study in critical size bone defect (rabbit model) showed complete bone repair in animals implanted with PHBHHx/SMC foams with and without rhBMP2, as determined by the formation of mature cortical bone and complete coverage of marrow cavity with both erythrocytes' medulla ossium rubra and medulla ossium flava after 8 weeks of implantation (Yang S, 2014).

6.5.2 Biocompatible Nanocomposite Foams with Organic Fillers

There have been growing interest in using biocompatible polysaccharide(nano)particles (such as cellulose nanocrystals (CNCs) or whiskers) and polymers (such as PEG) as fillers to reinforce nanocomposite foams for biomedical applications (Choi et al., 2021, Ju et al., 2020, Jafari et al., 2018, Ju et al., 2019, Lee et al., 2020, Liu et al., 2019, Mi et al., 2014, Padash et al., 2021, Shahrousvand et al., 2017, Wang et al., 2017). The high aspect ratio of CNCs has been shown to promote nucleation in microcellular injection molding. For example, Mi et al. fabricated PCL/ CNC nanocomposite foams by combining microcellular injection molding and ScCO$_2$ methods (Mi et al., 2014). They observed the reinforcing behavior of CNC (synthesized from pulp) causing an increase in foam tensile modulus with increasing CNC content. However, a reverse trend was noticed for pore size which reduced almost linearly with increasing CNC loading despite a marginal increase in total porosity. PCL/ CNC foams exhibited mixed biocompatibility in NIH/3T3 fibroblasts with high cell viability observed at lower CNC loading (0.5 and 1 wt%) with cells showing characteristic spindle shape fibroblast morphology. Cell viability dropped at 5 wt% CNC loading with cells exhibiting rounded cell morphology under SEM. The reduction in biocompatibility at high CNC content (5 wt%) was attributed to the sulfonyl hydroxide groups in CNCs (Mi et al., 2014). Ju et al. fabricated bimodal open-pore nanocomposite foam using poly(butylene succinate) (PBS) and CNC by modifying the ScCO$_2$ method (changing intermediate pressure) (Ju et al., 2020). The reinforcing behavior of CNC was only observed in bimodal porous PBS/CNC foam when compared to the unimodal porous foam. Bimodal porous PBS/CNC foam exhibited superior cell viability compared to unimodal porous PBS/CNC and neat PBS foams. The superior cell viability was assigned to the improved cell adhesion on bimodal porous foams probably due to a greater number of anchoring sites compared to unimodal porous foam, and higher hydrophilicity in comparison to neat PBS foam (Ju et al., 2020). In another study, biodegradable poly(3-hydroxybutyrate) (PHB) was supplemented with CNC to prepare a nanocomposite foam (PHB/CNC) using the non-solvent-induced phase separation method with subsequent freeze casting step (Choi et al., 2021). The authors dissolved PHB and CNC mixture in two different solvents THF and dioxane, respectively before adding these (PHB/CNC in THF and dioxane) solutions independently to non-solvent chloroform to induce phase polymer phase separation and gelation. The gels were frozen cast to obtain the nanocomposite foams. The two solvent systems (THF and dioxane) yielded PHB/CNC foams with completely different properties, based on the extent of the solubility of PHB and dispersion of CNC in each solvent – both of which (PHB solubility and CNC dispersion) were better in dioxane than in THF. Dioxane-based foams were stronger than THF foams despite similar overall porosity due to the more homogeneous distribution of CNC. SEM images revealed ECM-like morphology of the nanocomposite foams with THF-foams composed of thicker nanofiber walls than dioxane foams. Biocompatibility studies revealed cell viability of ~100% in human embryonic kidney 293T (HEK 293T) and neuronal model rat pheochromocytoma 12 (PC12) cells 24 h after incubation in THF-based neat PHB, dioxane-based neat PHB and dioxane-PHB/CNC foams. However, HEK 293T cell viability was reduced to ~90% in all three foams by day 4 of incubation. The difference in cell viability was more prominent in PC12 cells, where cell viability was lowest in dioxane-PHB/CNC foam followed by dioxane-PHB and THF-PHB foams by day 4 (Choi et al., 2021). The lower biocompatibility of dioxane foams could be due to the intrinsic toxicity of residual dioxane remaining in the foams.

Wang et al. prepared all polysaccharide-based nanocomposite foam by combining chitosan with nanofibrillated cellulose (NFC) from plants using the freeze-casting method (Wang et al., 2017). It was anticipated that greater intermolecular interactions (hydrogen bonding and electrostatic interactions) can be obtained between the filler and matrix by using a polysaccharide (as a matrix) instead of a synthetic polymer. They obtained porous foams with interconnected cell walls, the roughness of which increased with increasing NFC loading. The reinforcing effect of NFC was similar to CNC (reported in the above studies) where mechanical strength initially increased with increasing NFC content reaching the maximum value at 45 wt% loading but decreased with further increase in NFC loading. The chitosan/NFC foam exhibited a similar cell proliferation rate as TCPS over 96 h in L929 mouse fibroblasts cells. SEM images revealed foam–cell interaction with L929 cells extending on the foams into their characteristic spread morphology by anchoring to the NFC-induced rough edges on the nanocomposite foam surface (Wang et al., 2017).

In another study, PCL was reinforced with phospho-calcified cellulose nanowhiskers (PCCNWs) using the sugar and salt template method to form a multimodal porous foam (Jafari et al., 2018). PCCNWs were prepared by extracting CNW from wastepaper and calcifying it to neutralize the anionic phosphate groups to (a) improve their dispersion in PCL, and (b) utilize the calcium moiety to induce osteogenesis in hMSCs. PCCNWs had mixed responses on PCL foam properties – they (i) reduced the overall crystallinity of PCL via hydrogen bonding interaction between PCL carbonyl groups and PCCNW hydroxyl groups thus, restricting the movement of PCL chains; (ii) increased the overall strength of the nanocomposite foams with increasing PCCNW content but at the same time make them brittle. The nanocomposite foams were highly biocompatible with fibroblasts (SNL76/7) cells with cell viability of ~100% at all concentrations of PCCNWs. PCL/PCCNW foams induced osteogenic differentiation of hMSCs with significantly upregulated expression of ALP and calcium mineralization, both of which increased with increasing PCCNW loading indicating PCCNWs supported osteoblastic differentiation of hMSC (Jafari et al., 2018). This study highlights the potential of modified CNWs as an alternative to HA in bone regeneration applications.

Recently, PU-reinforced CNW foams have been explored for bone regeneration application (Shahrousvand et al., 2017, Padash et al., 2021). The PU/CNW foams with bimodal porosity were prepared using the sugar and salt template method. Similar to PCL foams, CNW forms hydrogen bonds with PU resulting in a significant enhancement in mechanical properties as reported by (Shahrousvand et al., 2017). They reported a marked increase in the tensile strength of the nanocomposite foam from 79 to 112 kPa with the addition of just 0.1 wt% CNW. The PU/CNW foams were highly biocompatible with hMSCs viability >90% over the 6 days study period at all CNW loadings. The osteogenic potential of the PU/CNW foam was observed with a significant increase in ALP expression and calcium mineralization over the 2 weeks incubation period compared to the TCPS control. More recently, in a modified approach, PU was supplemented with polyacrylonitrile (PAN) before loading with CNW to improve the biodegradability of the nanocomposite (PU/PAN/CNW) foams (Padash et al., 2021). The inclusion of PAN increased the hydrophilicity of the nanocomposite foams as observed with a reduction in water contact angle. The extent of hydrophilicity is considered a proxy for the biodegradability of a matrix. PU/PAN/CNW foams were highly biocompatible with cell viability of ~90% and induced significantly higher osteogenic response with upregulated calcium mineralization and osteogenic marker gene expressions (alkaline phosphatase, osteonectin, osteocalcin, collagen I and Runx 2) than the PU/PAN foam and TCPS controls (Padash et al., 2021). In an alternate strategy, PEG has been blended with the polymer matrix before the foaming process to improve the hydrophilicity and degradation of a nanocomposite foam. This strategy also enables to study the effect of solubility and diffusivity of Sc-CO_2 in different (individual) polymers within a blend on nanocomposite foam properties. For example, Liu et al. prepared PVA/PEG blended foams with bimodal cellular structure using the Sc-CO_2 method taking advantage of different solubilities of Sc-CO_2 in PVA and PEG (Liu et al., 2019). Different solubility of Sc-CO_2 led to smaller cells in the PEG phase and larger cells in the PVA phase. The surface morphology of the foam was dependent on PEG loading with a relatively smoother surface obtained at low PEG concentration, which became progressively rougher with increasing PEG loading. The change in surface (roughness) or pore morphologies (pore size) had a marginal effect on cellular response in mouse fibroblast (L929) cells, which maintained cell viability ~85–90%, exhibited strap-like shape and completely covered the foam surface over the 3 days incubation period (Liu et al., 2019). Ju and co-workers conducted a more comprehensive in vitro and in vivo bone regeneration study by blending PLLA with PEG. PLLA/PEG blended foams were prepared by the Sc-CO_2 method (Ju et al., 2019). They also observed an increase in surface roughness with increasing PEG loading, however, pore morphologies were completely different from the PVA/PEG foams prepared by Liu et al. (2019). Compared to the stretched pores in PVA/PEG foams, PLLA/PEG foams had a more uniform open-pore morphology similar to PU foams. Blended PLLA/PEG foam was highly biocompatible with MEF cell viability of ~95% compared to 90% in pure PLLA after 10 days in culture. In vivo rabbit bone defect model revealed mature osteoid tissues in animals treated with PLLA/PEG (5 wt%) foam three months are implantation. The implanted PLLA/PEG (5 wt%) foam also exhibited excellent cellular biocompatibility with significantly upregulated expression of Col1, BMP4, RUNX2 and BSP genes indicating bone formation compared to PLLA foam control (Liu et al., 2019).

6.5.3 BIOCOMPATIBLE NANOCOMPOSITE FOAMS FOR COMBINATORIAL BONE REGENERATION AND ANTIMICROBIAL APPLICATIONS

Nanocomposite foams have been explored for antimicrobial application and in some cases combinatorial antimicrobial and bone regeneration applications (Bužarovska et al., 2019, Mehrabani et al., 2018, Wang et al., 2014, Zhao et al., 2019). In regard to antimicrobial activity, fillers are typically selected with intrinsic antimicrobial properties such as silver (Ag) nanoparticles, graphene derivatives, transition metal dichalcogenides, TiO_2 and ZnO. The mechanism of action of these antimicrobial nanoparticles is well established, which primarily functions via the generation of reactive oxygen species (ROS) to promote oxidative stress-induced cell death in microbial cells.

Zhao et al. prepared water-soluble PU (WPU)/Ag nanoparticle nanocomposite antibacterial foams using the mechanical foaming method (Zhao et al., 2019). The Ag nanoparticles induced dose-dependent increase in mean porosity caused by foam cell coalescence resulting in increased pore interconnectivity. The antibacterial activity was determined using the colony counting method in both gram-negative (*Escherichia coli*) and gram-positive (*Staphylococcus aureus*) bacteria after incubating the bacterial cultures in foam dispersions. They observed bacteriostatic rates of 98.23% and 97.38% against *E. coli* and *S. aureus*, respectively in WPU/Ag (2 wt%) foams (Zhao et al., 2019). Wang et al. fabricated bacterial cellulose/ZnO nanocomposite foams by first preparing bacterial cellulose aerogel using the freeze-drying method and subsequently growing antimicrobial ZnO crystals using the solvothermal crystallization method (Wang et al., 2014). The ZnO loading was controlled by solvothermal crystallization reaction time. The cellulose/ZnO nanocomposite foams exhibited a unique hierarchical architecture with ZnO nanosphere-coated cellulose fibers in a string-beaded morphology. The antibacterial efficacy of the nanocomposite foams was studied in *E. coli* (strain DH5α) using the colony-forming unit counting method. The antibacterial activity was dependent on ZnO loading which improved with an increasing amount of ZnO coating reaching the maximum reduction in bacterial cell viability of ~98% after 5 h incubation under UV irradiation (at 365 nm) at 70 wt% ZnO loading. The antibacterial activities were significantly lower in dark conditions at the same ZnO loading, which points towards the involvement of ROS-dependent pathway (not determined experimentally). It was reasoned that the increase in antibacterial activity with ZnO loading was likely due to the increase in the total surface area and overall ZnO nanosphere size, which would lead to enhanced ROS production (Wang et al., 2014).

Multiple studies reported the development of multimodal and multifunctional nanocomposite foams as wound dressing material with additional antimicrobial effectiveness (Bužarovska et al., 2019, Mehrabani et al., 2018). Mehrabani et al. prepared a chitin/silk fibroin/TiO$_2$ nanocomposite foam as wound dressing material using the freeze casting method (Mehrabani et al., 2018). Chitin and silk fibroin were selected for their proven biocompatibility and biodegradability, and TiO$_2$ nanoparticles for their antimicrobial activity. Biocompatibility study in normal human dermal fibroblasts (HFFF2) cells revealed low cell viability of ~30–40%, which (cell viability) showed dose-dependent reduction with increasing concentration of TiO$_2$ in the nanocomposite foam. The antibacterial and antifungal activity was studied in *S. aureus*, *E. coli* and *C. albicans* (fungal strain) using the colony-forming unit counting method. They observed TiO$_2$ dose-dependent antimicrobial response in all three cultures with the highest activity recorded at 3 wt% TiO$_2$, which was significantly higher than control chitin/silk fibroin foam and TCPS (Mehrabani et al., 2018). It was proposed that TiO$_2$ nanoparticles released from the nanocomposite foam adhere to the bacterial cell wall causing morphological changes (in the cell wall) and interrupting material transfer across the cell wall inducing cell death. A hemolytic study revealed that TiO$_2$ nanoparticles had no impact on clotting ability with no difference in hemolysis observed between the chitin/silk fibroin/TiO$_2$ nanocomposite foam and pure chitin/silk fibroin foam control.

In another approach, Bužarovska and co-workers fabricated thermoplastic PU (TPU)/ZnO nanocomposite foam using the TIPS method (Bužarovska et al., 2019). Biocompatibility study in human adipose-derived stem cells (hASC) revealed the highest cell proliferation for 2 and 5 wt% ZnO nanocomposite foams compared to neat TPU foam a week after incubation. However, hASC proliferation was reduced with a further increase in ZnO content to 10 wt% in the nanocomposite foam. The antibacterial activity was studied in gram-positive (*Enterococcus faecalis and S. aureus*) and gram-negative (*E. coli and Pseudomonas aeruginosa*) monospecific (single cell type) biofilm formation assay. They observed a significant reduction in biofilm formation in all bacterial cultures in TPU/ZnO foams with the best response (maximum reduction) obtained at 10 wt% ZnO loading. It is clear that antimicrobial filler type and concentration should be carefully considered when designing a nanocomposite foam because of their (filler) potential toxicity towards mammalian cultures at higher concentrations and low antimicrobial efficacy at low concentrations.

6.6   CONCLUSION AND OUTLOOK

Research in nanocomposites has exploded in the last two to three decades fueled by the significant advancements made in the fields of polymer science, engineering, and nanotechnology. Nanocomposites in particular based on polymer/nanofillers provide access to the best of both worlds – the malleability of the polymers and the versatility of the nanofillers. Despite decades of research in nanocomposite design, fabrication and applications, interest in nanocomposite foams has just started to emerge. Even though polymer foams have penetrated every aspect of human life, nanocomposite foams are still relatively in their infancy. In particular, their applicability in the biomedical field remains grossly underexplored.

Most of the work reported to date on the biomedical application of nanocomposite foams focused primarily on establishing their biocompatibility in an in vitro model. In terms of disease models, the majority of the studies have been conducted on bone regeneration using stem cells and in a limited number of studies in vivo critical-size bone defects with promising results. However, in vivo testing of nanocomposite foams is still very premature and requires targeted research if these foams need to be translated into clinical applications. In vivo studies are also necessary to confirm and validate foam stability during implantation surgery and ambulatory movement of the animal for prolonged periods. Retaining the stability and shape of the nanocomposite foam during implantation surgery should be carefully considered. To this end, in situ fabrication of the foam at the injury site during surgery using the gas foaming method or the use of a shape recovery foam offers promising strategies.

It is evident that there are numerous synthetic and fabrication strategies available to prepare nanocomposite foams for biomedical applications. However, control over the intrinsic properties of the nanocomposite foam in terms of pore arrangement, surface features and distribution of the filler within the polymer network remains a challenge. Despite some initial promising studies as highlighted in this chapter, there is still a great need to conduct studies to compare different foam fabrication strategies with the same polymer and filler type and compositions to elucidate their utility for specific biomedical applications. Furthermore, there is no universal fabrication strategy that can tolerate nanofillers with different intrinsic properties (hydrophobic/hydrophilic, shape, size). The use of multiple fillers within a nanocomposite foam for biomedical applications is another avenue warranting further research. It is clear that the scope of research towards the biomedical application of nanocomposite foams is quite wide and there are different avenues that need exploring. It is envisaged that there is a bright future in biomedical nanocomposite foam research.

REFERENCES

Agarwal, V. & Chatterjee, K. 2018. Recent advances in the field of transition metal dichalcogenides for biomedical applications. *Nanoscale,* 10, 16365–16397.

Agarwal, V., Fadil, Y., Wan, A., Maslekar, N., Tran, B. N., Mat NOOR, R. A., Bhattacharyya, S., Biazik, J., Lim, S. & Zetterlund, P. B. 2021. Influence of anionic surfactants on the fundamental properties of polymer/reduced graphene oxide nanocomposite films. *ACS Applied Materials & Interfaces,* 13, 18338–18347.

Agarwal, V., Varghese, N., Dasgupta, S., Sood, A. K. & Chatterjee, K. 2019. Engineering a 3D MoS2 foam using keratin exfoliated nanosheets. *Chemical Engineering Journal,* 374, 254–262.

Agarwal, V., Wood, F. M., Fear, M. & Iyer, K. S. 2017. Polymeric nanofibre Scaffold for the delivery of a transforming growth factor β1 inhibitor. *Australian Journal of Chemistry,* 70, 280–285.

Armentano, I., Dottori, M., Fortunati, E., Mattioli, S. & Kenny, J. M. 2010. Biodegradable polymer matrix nanocomposites for tissue engineering: A review. *Polymer Degradation and Stability,* 95, 2126–2146.

Aslam Khan, M. U., Raza, M. A., Mehboob, H., Abdul Kadir, M. R., Abd Razak, S. I., Shah, S. A., Iqbal, M. Z. & Amin, R. 2020. Development and in vitro evaluation of κ-carrageenan based polymeric hybrid nanocomposite scaffolds for bone tissue engineering. *RSC Advances,* 10, 40529–40542.

Boccaccini, A. R., Erol, M., Stark, W. J., Mohn, D., Hong, Z. & Mano, J. F. 2010. Polymer/bio-active glass nanocomposites for biomedical applications: A review. *Composites Science and Technology,* 70, 1764–1776.

Bužarovska, A., Dinescu, S., Lazar, A. D., Serban, M., Pircalabioru, G. G., Costache, M., Gualandi, C. & Avérous, L. 2019. Nanocomposite foams based on flexible biobased thermoplastic polyurethane and ZnO nanoparticles as potential wound dressing materials. *Materials Science and Engineering: C,* 104, 109893.

Carfì Pavia, F., Conoscenti, G., Greco, S., La Carrubba, V., Ghersi, G. & Brucato, V. 2018. Preparation, characterization and in vitro test of composites poly-lactic acid/hydroxy-apatite scaffolds for bone tissue engineering. *International Journal of Biological Macromolecules,* 119, 945–953.

Chen, C., Garber, L., Smoak, M., Fargason, C., Scherr, T., Blackburn, C., Bacchus, S., Lopez, M. J., Pojman, J. A., Del Piero, F. & Hayes, D. J. 2015. In vitro and in vivo character-ization of pentaerythritol triacrylate-co-trimethylolpropane nanocomposite scaffolds as potential bone augments and grafts. *Tissue Engineering Part A,* 21, 320–331.

Chen, L., Rende, D., Schadler, L. S. & Ozisik, R. 2013. Polymer nanocomposite foams. *Journal of Materials Chemistry A,* 1, 3837–3850.

Choi, J., Kang, J. & Yun, S. I. 2021. Nanofibrous foams of poly(3-hydroxybutyrate)/cellulose nanocrystal composite fabricated using non-solvent-induced phase separation. *Langmuir,* 37, 1173–1182.

Delafresnaye, L., Dugas, P.-Y., Dufils, P.-E., Chaduc, I., Vinas, J., Lansalot, M. & Bourgeat-Lami, E. 2017. Synthesis of clay-armored poly(vinylidene chloride-co-methyl acrylate) latexes by Pickering emulsion polymerization and their film-forming properties. *Polymer Chemistry,* 8, 6217–6232.

Derakhshanfar, S., Mbeleck, R., Xu, K., Zhang, X., Zhong, W. & Xing, M. 2018. 3D bio-printing for biomedical devices and tissue engineering: A review of recent trends and advances. *Bioactive Materials,* 3, 144–156.

Deville, S. 2017. *Freezing colloids: observations, principles, control, and use: Applications in materials science, life science, earth science, food science, and engineering*, Springer.

Eroglu, E., Agarwal, V., Bradshaw, M., Chen, X., Smith, S. M., Raston, C. L. & Swaminathan Iyer, K. 2012. Nitrate removal from liquid effluents using microalgae immobilized on chitosan nanofiber mats. *Green Chemistry*, 14, 2682–2685.

Eryildiz, M. & Altan, M. 2020. Fabrication of polylactic acid/halloysite nanotube scaffolds by foam injection molding for tissue engineering. *Polymer Composites*, 41, 757–767.

Fadil, Y., Dinh, L. N. M., Yap, M. O. Y., Kuchel, R. P., Yao, Y., Omura, T., Aregueta-Robles, U. A., Song, N., Huang, S., Jasinski, F., Thickett, S. C., Minami, H., Agarwal, V. & Zetterlund, P. B. 2019. Ambient-temperature waterborne polymer/rGO nanocomposite films: Effect of rGO distribution on electrical conductivity. *ACS Applied Materials & Interfaces*, 11, 48450–48458.

Fadil, Y., Thickett, S. C., Agarwal, V. & Zetterlund, P. B. 2022. Synthesis of graphene-based polymeric nanocomposites using emulsion techniques. *Progress in Polymer Science*, 125, 101476.

Garber, L., Chen, C., Kilchrist, K. V., Bounds, C., Pojman, J. A. & Hayes, D. 2013. Thiol-acrylate nanocomposite foams for critical size bone defect repair: A novel biomaterial. *Journal of Biomedical Materials Research Part A*, 101, 3531–3541.

He, J., Zhang, Z., Xiao, C., Liu, F., Sun, H., Zhu, Z., Liang, W. & Li, A. 2020. High-performance salt-rejecting and cost-effective superhydrophilic porous monolithic polymer foam for solar steam generation. *ACS Applied Materials & Interfaces*, 12, 16308–16318.

Ho, D., Zou, J., Chen, X., Munshi, A., Smith, N. M., Agarwal, V., Hodgetts, S. I., Plant, G. W., Bakker, A. J., Harvey, A. R., Luzinov, I. & Iyer, K. S. 2015. Hierarchical patterning of multifunctional conducting polymer nanoparticles as a bionic platform for topographic contact guidance. *ACS Nano*, 9, 1767–1774.

Huang, Z., Wan, Y., Zhu, X., Zhang, P., Yang, Z., Yao, F. & Luo, H. 2021. Simultaneous engineering of nanofillers and patterned surface macropores of graphene/hydroxyapatite/polyetheretherketone ternary composites for potential bone implants. *Materials Science and Engineering: C*, 123, 111967.

Jafari, H., Shahrousvand, M. & Kaffashi, B. 2018. Reinforced Poly(ε-caprolactone) bimodal foams via phospho-calcified cellulose nanowhisker for osteogenic differentiation of human mesenchymal stem cells. *ACS Biomaterials Science & Engineering*, 4, 2484–2493.

Ju, J., Gu, Z., Liu, X., Zhang, S., Peng, X. & Kuang, T. 2020. Fabrication of bimodal open-porous poly (butylene succinate)/cellulose nanocrystals composite scaffolds for tissue engineering application. *International Journal of Biological Macromolecules*, 147, 1164–1173.

Ju, J., Peng, X., Huang, K., Li, L., Liu, X., Chitrakar, C., Chang, L., Gu, Z. & Kuang, T. 2019. High-performance porous PLLA-based scaffolds for bone tissue engineering: Preparation, characterization, and in vitro and in vivo evaluation. *Polymer*, 180, 121707.

Karadas, O., Yucel, D., Kenar, H., Torun Kose, G. & Hasirci, V. 2014. Collagen scaffolds with in situ-grown calcium phosphate for osteogenic differentiation of Wharton's jelly and menstrual blood stem cells. *Journal of Tissue Engineering and Regenerative Medicine* 8, 534–545.

Karakoti, A. S., Tsigkou, O., Yue, S., Lee, P. D., Stevens, M. M., Jones, J. R. & Seal, S. 2010. Rare earth oxides as nanoadditives in 3-D nanocomposite scaffolds for bone regeneration. *Journal of Materials Chemistry*, 20, 8912–8919.

Kausar, A. 2020. Thermally conducting polymer/nanocarbon and polymer/inorganic nanoparticle nanocomposite: A review. *Polymer-Plastics Technology and Materials*, 59, 895–909.

Kishan, A. P. & Cosgriff-Hernandez, E. M. 2017. Recent advancements in electrospinning design for tissue engineering applications: A review. *Journal of Biomedical Materials Research Part A*, 105, 2892–2905.

Lee, L. J., Zeng, C., Cao, X., Han, X., Shen, J. & Xu, G. 2005. Polymer nanocomposite foams. *Composites Science and Technology,* 65, 2344–2363.

Lee, R. E., Azdast, T., Wang, G., Wang, X., Lee, P. C. & Park, C. B. 2020. Highly expanded fine-cell foam of polylactide/polyhydroxyalkanoate/nano-fibrillated polytetrafluoroethylene composites blown with mold-opening injection molding. *International Journal of Biological Macromolecules,* 155, 286–292.

Li, B., Agarwal, V., Ho, D., Vede, J.-P. & Iyer, K. S. 2018. Systematic assessment of surface functionality on nanoscale patterns for topographic contact guidance of cells. *New Journal of Chemistry,* 42, 7237–7240.

Li, Y., Huang, X., Zeng, L., Li, R., Tian, H., Fu, X., Wang, Y. & Zhong, W.-H. 2019. A review of the electrical and mechanical properties of carbon nanofiller-reinforced polymer composites. *Journal of Materials Science,* 54, 1036–1076.

Liang, D., Hsiao, B. S. & Chu, B. 2007. Functional electro spun nanofibrous scaffolds for biomedical applications. *Advanced Drug Delivery Reviews,* 59, 1392–1412.

Liao, X., Zhang, H. & He, T. 2012. Preparation of porous biodegradable polymer and its nanocomposites by supercritical CO_2 foaming for tissue engineering. *Journal of Nanomaterials,* 2012, 836394.

Liu, P., Chen, W., Liu, C., Tian, M. & Liu, P. 2019. A novel poly (vinyl alcohol)/poly (ethylene glycol) scaffold for tissue engineering with a unique bimodal open-celled structure fabricated using supercritical fluid foaming. *Scientific Reports,* 9, 9534.

Lundin, J. G., McGann, C. L., Daniels, G. C., Streifel, B. C. & Wynne, J. H. 2017. Hemostatic kaolin-polyurethane foam composites for multifunctional wound dressing applications. *Materials Science and Engineering: C,* 79, 702–709.

Mallakpour, S. & Rashidimoghadam, S. 2017. 8- Recent developments in the synthesis of hybrid polymer/clay nanocomposites: Properties and applications. *In:* Thakur, V. K., Thakur, M. K. & Pappu, A. (eds.) *Hybrid Polymer Composite Materials.* Woodhead Publishing.

Manoukian, O. S., Arul, M. R., Rudraiah, S., Kalajzic, I. & Kumbar, S. G. 2019. Aligned microchannel polymer-nanotube composites for peripheral nerve regeneration: Small molecule drug delivery. *Journal of Controlled Release,* 296, 54–67.

Maslekar, N., Mat Noor, R. A., Kuchel, R. P., Yao, Y., Zetterlund, P. B. & Agarwal, V. 2020. Synthesis of diamine functionalised graphene oxide and its application in the fabrication of electrically conducting reduced graphene oxide/polymer nanocomposite films. *Nanoscale Advances,* 2, 4702–4712.

Mehrabani, M. G., Karimian, R., Rakhshaei, R., Pakdel, F., Eslami, H., Fakhrzadeh, V., Rahimi, M., Salehi, R. & Kafil, H. S. 2018. Chitin/silk fibroin/TiO2 bio-nanocomposite as a biocompatible wound dressing bandage with strong antimicrobial activity. *International Journal of Biological Macromolecules,* 116, 966–976.

Meka, S. R. K., Agarwal, V. & Chatterjee, K. 2019. In situ preparation of multicomponent polymer composite nanofibrous scaffolds with enhanced osteogenic and angiogenic activities. *Materials Science and Engineering: C,* 94, 565–579.

Meka, S. R. K., Kumar Verma, S., Agarwal, V. & Chatterjee, K. 2018. In situ silication of polymer nanofibers to engineer multi-biofunctional composites. *ChemistrySelect,* 3, 3762–3773.

Melchels, F. P. W., Feijen, J. & Grijpma, D. W. 2010. A review on stereolithography and its applications in biomedical engineering. *Biomaterials,* 31, 6121–6130.

Meng, Q., Han, S., Liu, T., Ma, J., Ji, S., Dai, J., Kang, H. & Ma, J. 2020. Noncovalent modification of boron nitrite nanosheets for thermally conductive, mechanically resilient epoxy nanocomposites. *Industrial & Engineering Chemistry Research,* 59, 20701–20710.

Mi, H.-Y., Jing, X., Peng, J., Salick, M. R., Peng, X.-F. & Turng, L.-S. 2014. Poly(ε-caprolactone) (PCL)/cellulose nano-crystal (CNC) nanocomposites and foams. *Cellulose,* 21, 2727–2741.

Mittal, V. 2014. Functional polymer nanocomposites with graphene: A review. *Macromolecular Materials and Engineering*, 299, 906–931.

Moghadas, B. K., Akbarzadeh, A., Azadi, M., Aghili, A., Rad, A. S. & Hallajian, S. 2020. The morphological properties and biocompatibility studies of synthesized nanocomposite foam from modified polyethersulfone/graphene oxide using supercritical CO2. *Journal of Macromolecular Science, Part A*, 57, 451–460.

Owen, R., Sherborne, C., Evans, R., Reilly, G. C. & Claeyssens, F. 2020. Combined porogen leaching and emulsion templating to produce bone tissue engineering Scaffolds. *International Journal of Bioprinting*, 6, 265–265.

Padash, A., Halabian, R., Salimi, A., Kazemi, N. M. & Shahrousvand, M. 2021. Osteogenic differentiation of mesenchymal stem cells on the bimodal polymer polyurethane/polyacrylonitrile containing cellulose phosphate nanowhisker. *Human Cell*, 34, 310–324.

Palchoudhury, S., Baalousha, M. & Lead, J. R. 2015. Chapter 5 – Methods for measuring concentration (mass, surface area and number) of nanomaterials. *In:* Baalousha, M. & Lead, J. R. (eds.) *Frontiers of Nanoscience*. Elsevier.

Panahi-Sarmad, M., Noroozi, M., Abrisham, M., Eghbalinia, S., Teimoury, F., Bahramian, A. R., Dehghan, P., Sadri, M. & Goodarzi, V. 2020. A comprehensive review on carbon-based polymer nanocomposite foams as electromagnetic interference shields and piezoresistive sensors. *ACS Applied Electronic Materials*, 2, 2318–2350.

Parai, R. & Bandyopadhyay-Ghosh, S. 2019. Engineered bio-nanocomposite magnesium scaffold for bone tissue regeneration. *Journal of the Mechanical Behavior of Biomedical Materials*, 96, 45–52.

Pelaseyed, S. S., Madaah Hosseini, H. R. & Samadikuchaksaraei, A. 2020. A novel pathway to produce biodegradable and bioactive PLGA/TiO2 nanocomposite scaffolds for tissue engineering: Air–liquid foaming. *Journal of Biomedical Materials Research*, 108, 1390–1407.

Pina, S., Oliveira, J. M. & Reis, R. L. 2015. Natural-based nanocomposites for bone tissue engineering and regenerative medicine: A review. *Advanced Materials*, 27, 1143–1169.

Qiu Li Loh & Choong, C. 2013. Three-dimensional Scaffolds for tissue engineering applications: Role of porosity and pore size. *Tissue Engineering Part B: Reviews*, 19, 485–502.

Salerno, A., Zeppetelli, S., Di Maio, E., Iannace, S. & Netti, P. A. 2011a. Design of bimodal PCL and PCL-HA nanocomposite Scaffolds by two step depressurization during solid-state supercritical CO2 foaming. *Macromolecular Rapid Communications*, 32, 1150–1156.

Salerno, A., Zeppetelli, S., Di Maio, E., Iannace, S. & Netti, P. A. 2011b. Processing/structure/property relationship of multi-scaled PCL and PCL–HA composite scaffolds prepared via gas foaming and NaCl reverse templating. *Biotechnology and Bioengineering*, 108, 963–976.

Samantaray, P. K., Indrakumar, S., Chatterjee, K., Agarwal, V. & Bose, S. 2020. 'Template-free' hierarchical MoS2 foam as a sustainable 'green' scavenger of heavy metals and bacteria in point of use water purification. *Nanoscale Advances*, 2, 2824–2834.

Scotti, K. L. & Dunand, D. C. 2018. Freeze casting–A review of processing, microstructure and properties via the open data repository, FreezeCasting.net. *Progress in Materials Science*, 94, 243–305.

Shahrousvand, E., Shahrousvand, M., Ghollasi, M., Seyedjafari, E., Jouibari, I. S., Babaei, A. & Salimi, A. 2017. Preparation and evaluation of polyurethane/cellulose nanowhisker bimodal foam nanocomposites for osteogenic differentiation of hMSCs. *Carbohydrate Polymers*, 171, 281–291.

Silverstein, M. S. & Cameron, N. R. 2010. PolyHIPEs—porous polymers from high internal phase emulsions. *Encyclopedia of Polymer Science and Technology*. DOI: 10.1002/0471440264.pst571

Singh, A. K., Shishkin, A., Koppel, T. & Gupta, N. 2018. A review of porous lightweight composite materials for electromagnetic interference shielding. *Composites Part B: Engineering*, 149, 188–197.

Stucki, M., Loepfe, M. & Stark, W. J. 2018. Porous polymer membranes by hard templating–A review. *Advanced Engineering Materials*, 20, 1700611.

Tran, B. N., Bhattacharyya, S., Yao, Y., Agarwal, V. & Zetterlund, P. B. 2021a. In situ surfactant effects on polymer/reduced graphene oxide nanocomposite films: Implications for coating and biomedical applications. *ACS Applied Nano Materials*, 4, 12461–12471.

Tran, B. N., Thickett, S. C., Agarwal, V. & Zetterlund, P. B. 2021b. Influence of polymer matrix on polymer/graphene oxide nanocomposite intrinsic properties. *ACS Applied Polymer Materials*, 3, 5145–5154.

Uddin, M. N., Dhanasekaran, P. S. & Asmatulu, R. 2019. Mechanical properties of highly porous PEEK bionanocomposites incorporated with carbon and hydroxyapatite nanoparticles for scaffold applications. *Progress in Biomaterials*, 8, 211–221.

Wang, J., Luo, J., Kunkel, R., Saha, M., Bohnstedt, B. N., Lee, C.-H. & Liu, Y. 2019. Development of shape memory polymer nanocomposite foam for treatment of intracranial aneurysms. *Materials Letters*, 250, 38–41.

Wang, L., Gong, C., Shen, Y., Xu, M., He, G., Wang, L. & Song, Y. 2018. Conjugated Schiff base polymer foam/macroporous carbon integrated electrode for electrochemical sensing. *Sensors and Actuators B: Chemical*, 265, 227–233.

Wang, P., Zhao, J., Xuan, R., Wang, Y., Zou, C., Zhang, Z., Wan, Y. & Xu, Y. 2014. Flexible and monolithic zinc oxide bionanocomposite foams by a bacterial cellulose mediated approach for antibacterial applications. *Dalton Transactions*, 43, 6762–6768.

Wang, Y., Uetani, K., Liu, S., Zhang, X., Wang, Y., Lu, P., Wei, T., Fan, Z., Shen, J., Yu, H., Li, S., Zhang, Q., Li, Q., Fan, J., Yang, N., Wang, Q., Liu, Y., Cao, J., Li, J. & Chen, W. 2017. Multifunctional bionanocomposite foams with a Chitosan matrix reinforced by nanofibrillated cellulose. *ChemNanoMat*, 3, 98–108.

Wu, S., Liu, X., Yeung, K. W. K., Liu, C. & Yang, X. 2014. Biomimetic porous scaffolds for bone tissue engineering. *Materials Science and Engineering: R: Reports*, 80, 1–36.

Xie, R., Hu, J., Ng, F., Tan, L., Qin, T., Zhang, M. & Guo, X. 2017. High performance shape memory foams with isocyanate-modified hydroxyapatite nanoparticles for minimally invasive bone regeneration. *Ceramics International*, 43, 4794–4802.

Yang, J., Yang, W., Chen, W. & Tao, X. 2020. An elegant coupling: Freeze-casting and versatile polymer composites. *Progress in Polymer Science*, 109, 101289.

Yang, S., Xu, S., Zhou, P, Wang, J., Tan, H., Liu, Y., Tang, T., Liu, C. & Huang, L. 2014. Siliceous mesostructured cellular foams/poly(3-hydroxybutyrate-co-3-hydroxyhexanoate) composite biomaterials for bone regeneration. *International Journal of Nanomedicine*, 9, 4795–4807.

Zabihi, O., Ahmadi, M., Nikafshar, S., Chandrakumar Preyeswary, K. & Naebe, M. 2018. A technical review on epoxy-clay nanocomposites: Structure, properties, and their applications in fiber reinforced composites. *Composites Part B: Engineering*, 135, 1–24.

Zare, Y. & Shabani, I. 2016. Polymer/metal nanocomposites for biomedical applications. *Materials Science and Engineering: C*, 60, 195–203.

Zeltinger, J., Sherwood, J. K., Graham, D. A., Müeller, R. & Griffith, L. G. 2001. Effect of pore size and void fraction on cellular adhesion, proliferation, and matrix deposition. *Tissue Engineering*, 7, 557–572.

Zhao, B., Qian, Y., Qian, X., Fan, J. & Feng, Y. 2019. Fabrication and characterization of waterborne polyurethane/silver nanocomposite foams. *Polymer Composites*, 40, 1492–1498.

Ziminska, M., Chalanqui, M. J., Chambers, P., Acheson, J. G., Mccarthy, H. O., DUNNE, N. J. & Hamilton, A. R. 2019. Nanocomposite-coated porous templates for engineered bone scaffolds: a parametric study of layer-by-layer assembly conditions. *Biomedical Materials*, 14, 065008.

Zuev, D. M., Nguyen, A. K., Putlyaev, V. I. & Narayan, R. J. 2020. 3D printing and bioprinting using multiphoton lithography. *Bioprinting*, 20, e00090.

7 Polymer Nanocomposite Foams and Acoustics

Benjamin Tawiah, Charles Frimpong and Bismark Sarkodie

CONTENTS

7.1 INTRODUCTION

Various materials have been modified to satisfy the insatiable desire for specialized products with versatile applications in various fields. As a result, many different methods and nanomaterials with unique properties have been used to enhance the electrical, mechanical, flame retardant, electromagnetic interference shielding, catalytic, and other properties of conventional polymers to help meet stringent product requirements and ensure environmental safety. Among these products are polymers, polymer nanocomposites, and their derivatives, such as polymer nanocomposite foams. Parameters such as the temperature involved in the fabrication of these products affect their outcome. Polymers like poly(aniline), polyvinylpyrrolidone, polypyrrole, poly (3,4-ethylenedioxythiophene), and a variety with the component polystyrene sulfonate, as well as fluoride-based polymers like polyvinylidene difluoride, and poly (vinylidene fluoride-trifluoroethylene) are widely used. Due to the limitations of conventional polymers, such as heat instability, which restricts their use,

DOI: 10.1201/9781003218692-7

elastomer reinforcement using nanofillers has become imperative. For example, elastomers lack strong intermolecular interaction and crystallinity, hence nanofillers are used to improve their mechanical properties considerably. Furthermore, conductive nanofillers can enhance the anti-static efficiency, heat capacity, gas barrier effect, and fire retardant performance of elastomeric foams. Polymer nanocomposites have been fabricated in various forms, sizes, and structures to increase their applications in diverse fields. Foaming has emerged as an innovative technique to enhance the desirable properties of polymer nanocomposites, due to the creation of cell structures and their low weight, in the middle of structural modification of polymer nanocomposites.

Because of increasing urbanization and the rising demand for energy in the 21st century, sound pollution has recently become a danger. The desire for home safety and a healthier environment has grown to be a significant requirement for communities. Several investigations on sound-absorbing materials have been conducted to find a solution to these difficulties. Polymer nanocomposite foams are used extensively in acoustic engineering applications for industrial noise control, room acoustics, automobile acoustics, and studio acoustics. Hence, the desire for improved sound absorption materials and ways to accurately evaluate their acoustic capacity has become an important research area in recent years.

This chapter introduces polymer nanocomposite and reveals the main distinguishing features of polymer nanocomposite compared to solid polymer nanocomposite. Some major fabrication methods, blowing agents, and other parameters that influence the structure and properties of the cells in the polymer nanocomposite foams have been discussed.

7.2 POLYMER NANOCOMPOSITES

When compared to their distinct components, polymer nanocomposites are a blend of polymer continuous phase and nanoparticles discontinuous phase that exhibit many mechanical, electrical, and optical characteristics. In other words, a multiphase solid material with one, two, or three-dimension materials of less than 100 nanometres, or materials with nano-scale repeating intervals amongst the multiple segments that form the structure can be referred to as *polymer nanocomposite*. In its wider context, the term "nanocomposite" can refer to porous media, colloids, gels, and copolymers, but it is most commonly used to define a solid mixture of a bulk matrix and nano-dimensional phase(s) with characteristics that differ owing to structural and chemical differences. The unique properties of polymer nanocomposites have piqued the interest of experts all over the world in recent years. Polymer nanocomposites may be produced in many nanoscale forms with minimum effort, allowing the fabrication of various new chemical and biological products (Das et al., 2018, Tyagi and Tyagi, 2014). Even though polymers may be utilized as structural materials without reinforcement, their application is limited due to their inadequate mechanical characteristics. Composite materials are a novel replacement for traditionally filled polymeric materials. Composite materials are materials in which nano inorganic materials, generally 10–100 nm in multiple dimensions (1D, 2D, and 3D), are distributed in an organic polymer matrix. When compared to plain polymers or conventional composites, polymer nanocomposites have significantly enhanced characteristics due to the nanoscale diameters of the nanofillers (Tyagi and Tyagi, 2014). In the automotive, aerospace, construction, and electronics sectors, polymer composites are widely utilized (Lee et al., 2005, 2012).

7.2.1 POLYMER NANOCOMPOSITE FOAMS

Polymeric foams, also known as cellular or expanded polymers, are a common part of life due to their lightweight and low compressive creep properties, which have made them appealing for diverse industrial and domestic uses (Gao et al., 2020). Polymer foams are a type of lightweight material that is used in a wide range of applications with a market value of over $2 billion annually (Wu et al., 2021, Xie et al., 2021). Foams are porous light materials with unique properties that are commonly used in the automotive and aerospace sectors. These industries demand lightweight yet stiff and mechanically robust materials since the velocity and security of such vehicles are connected to their mechanical features. Foams are progressively replacing the traditional metal and other kinds of composites due to their corrosion and creep tolerance, as well as certain mechanical properties. Furthermore, well-engineered foams may readily be employed in tolerance components and can also improve material flame and smoke characteristics (Yuceturk et al., 2018). Foams with poor mechanical properties, low wear resistance, and inherently low thermal and dimensional stability have limited applications in high-end engineering materials.

The Montreal Protocol has determined that the most frequently used chlorofluorocarbon blowing agents in the manufacturing of foams causes ozone depletion in the upper stratosphere, and therefore, have been prohibited since 2010 (Lee et al., 2005). The addition of nanofillers to create polymer nanocomposite foams has broadened the range of applications for polymer nanocomposite foams due to the extra qualities embedded by the nanofillers, such as increased thermal stability and electrical properties, among others (Arab-Baraghi et al., 2014, Gao et al., 2020, Han, 2015, Kuang et al., 2016, Sundarram et al., 2015, Wang et al., 2019, Xie et al., 2021). The porous structural nature of polymer nanocomposite foams has an advantage in regions where reflection and adsorption are important. In comparison to polymer nanocomposites, it has been demonstrated that the porous structure of polymer nanocomposite foams substantially increases the internal multiple reflections and prolongs the activities that lead to electromagnetic waves in foam equipment, particularly conductive polymer composite foams, making it ideal for fabricating absorption-dominated electromagnetic interference shielding materials (Duan et al., 2020). Microwaves entering the composite traverse the longest absorption and reflection path due to the heterogeneous filler network structure of polymer nanocomposite foams. Polymer foams with electromagnetic functional gradients presumably emit extremely low microwave transmittance and reflection (Duan et al., 2020).

When gas foaming is used to generate porous structures, closed and/or open cells are formed as shown in Figure 7.1. A limited number of well-distributed nanofillers in the polymer domain potentially act as a nucleation site, accelerating the bubble nucleation process. Several parameters must be accomplished to create a closed-cell foam. An important consideration is that closed-cell foam cannot have more than two

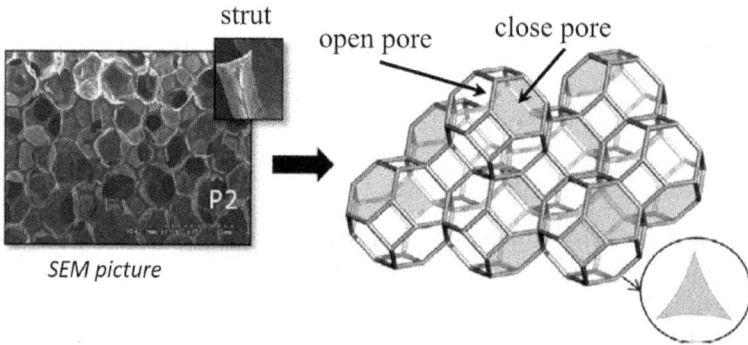

FIGURE 7.1 Cellular structure of polymer nanocomposite foam (Asadi Khanouki and Ohadi, 2018).

ruptured cell walls per cell unit area or aperture, and the cell framework ought to always be communal to at least three structural components within the cell. The various characteristics of foams are due to structural variations. Foams with a closed-cell structure are good insulators. Open-cell foams, in contrast to closed-cell foams, may absorb sound much better. Such foams have higher gas and steam permeability, but less thermal insulation and electrical conductivity. Mechanical or thermal treatment can be used to convert closed-cell materials into open-cell foams (Lee et al., 2005, Olszta et al., 2007, Sinha Ray, 2013, Standau and Altstädt, 2019).

7.2.2 METHODS OF POLYMER NANOCOMPOSITE FOAMS PREPARATION

Melt-blending, resin-transfer, solution blending, vacuum-assisted filtering, and other methods have been used to create polymer nanocomposite (Si et al., 2019, Song et al., 2021, Zhang et al., 2011). The foaming agent employed distinguishes the preparation of polymer nanocomposites from polymer nanocomposite foams. To fabricate multifunctional nanocomposite foams with low density by extrusion, injection moulding, freeze-drying, solid-state foaming, supercritical carbon dioxide ($ScCO_2$) and water vapour-induced dissociation may all be employed (Yuceturk et al., 2018). The different techniques of foam preparation are all regulated by the same foaming principles. The phases in the polymer foaming process are depicted in Figure 7.2. A gas is blended into a polymer phase, which is then homogenized into a single melt phase to make polymer foam. The sort of material, gas pressure, and temperature, all influence its solubility. When the blowing agent in the form of liquid or $ScCO_2$ is supplied to the melted polymer, fluid molecules are maintained at the surface of the polymer phase. The composition of the blowing agent (e.g., polarity, molecular size) influences the solubility, whereas the circulation of the blowing agent molecules in the polymer is affected by partial pressure or concentration gradients. The nucleation of the polymer melt, cell formation by multiply effect, enlargement, and foam stabilization is all part of the foaming process (Standau and Altstädt, 2019). In the manufacture of polymer nanocomposite foam, the percentage of nanofillers,

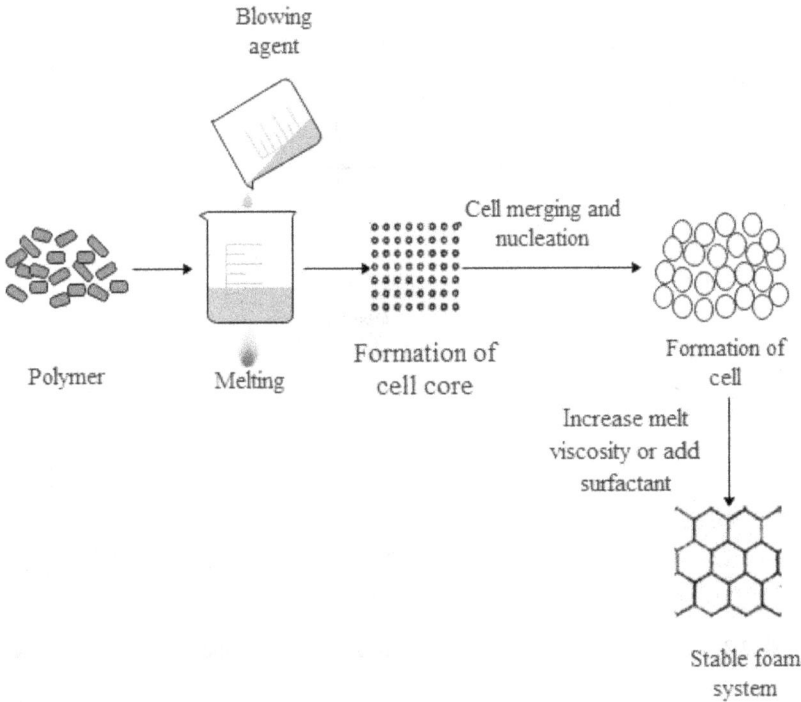

FIGURE 7.2   Schematic foaming process (Jin et al., 2019).

nanoparticle size, and shape as well as the orientation mixed with the polymer affects the final composite foam. To put it another way, the nucleation site separates polymer foaming from polymer nanocomposite manufacturing. In polymer nanocomposites, fillers are the major nucleation sites determining the characteristics of foams in the nanocomposite.

A batch technology that uses just a high-pressure gas dissolving technique to foam an initially synthesized polymer material in an autoclave reactor, or a system in which a previously extruded-composite material is foamed in an autoclave reaction chamber through a high-pressure gas dissolution technique to produce thermoplastic-based foam or direct extrusion in which foam is produced directly at the exit of an extrusion die, primarily by physical blowing agents such as CO_2 or N_2 (Ray, 2013).

Batch foaming is a one-time procedure that occurs in an autoclave at a constant temperature and pressure. This technique is simple to use and produces foams that also have a uniform structure and content. Temperature and pressure-induced batch foaming are the two forms of batch foaming techniques. The nanocomposite is first saturated with blowing agents in an autoclave in the temperature-induced batch foaming process, as illustrated in Figure 7.3. To avoid foaming, pressure is unrestrained at a low temperature once saturation has been attained. The supersaturated sample is then immersed in a heated oil bath for a predetermined amount of time to promote cell nucleation and growth. To stabilize the cell structure, the sample is

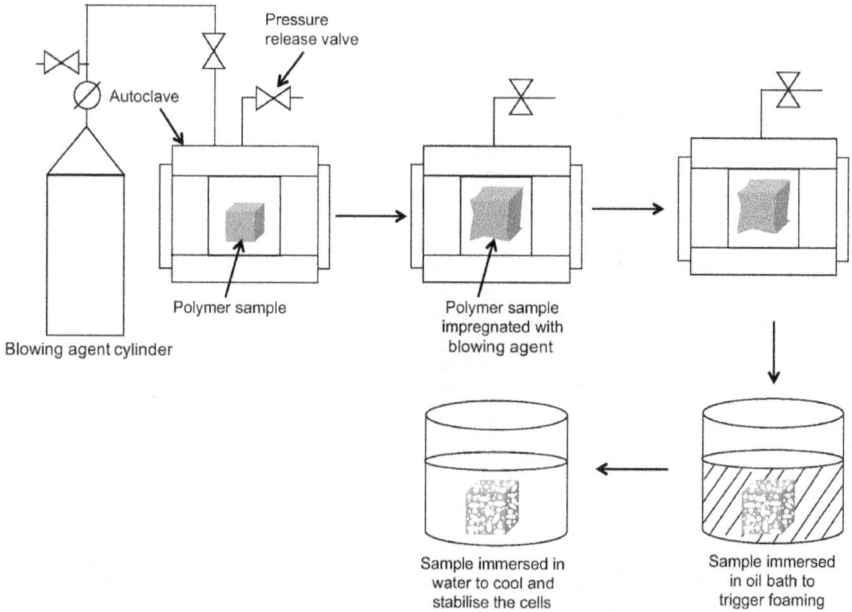

FIGURE 7.3 Temperature-induced batch foaming process (Okolieocha et al., 2015).

removed from the heated oil bath and submerged in a cooling bath containing water or oil (Standau and Altstädt, 2019, Fan et al., 2019).

As demonstrated in Figure 7.4, pressure-induced batch foaming comprises two steps: (a) sorption of blowing agents, and (b) depressurization. In this process, blowing agents are first liquefied and dispersed into the polymer nanocomposite matrix at a carefully regulated temperature and pressure. The pressure is promptly released to air pressure after the specimen has been inundated with blowing agents for a period. Cell nucleation and growth are aided by a decrease in pressure. Temperature, pressure, blowing agent type, blowing agent sorption time, pressure drop, and pressure drop rate are all factors that influence foam cell structure and foam expansion in this process. Foams with the appropriate cell morphology may be made by adjusting these processing parameters. It has been discovered that rapid decompression reduces cell sizes and increases cell density. A faster rate of depressurization causes more supersaturation and a larger mass transfer coefficient for cell nucleation. Because a faster depressurization rate reduces the time it takes for the blowing agent to diffuse, the more blowing agent may be necessary for cell initiation and formation rather than cell development, which may result in smaller cell sizes yet higher cell density (Standau and Altstädt, 2019, Okolieocha et al., 2015).

Foam injection moulding is done with a screw-driven injection moulding machine (Figure 7.5). A rectangular chamber is included in the mould, with a fan gate linking it to the sprue. Low-pressure foam injection moulding is often used for samples with a void fraction of around 30%, whereas high-pressure foam injection moulding IM + MO is required for samples with greater void fractions. The polymer nanocomposite/

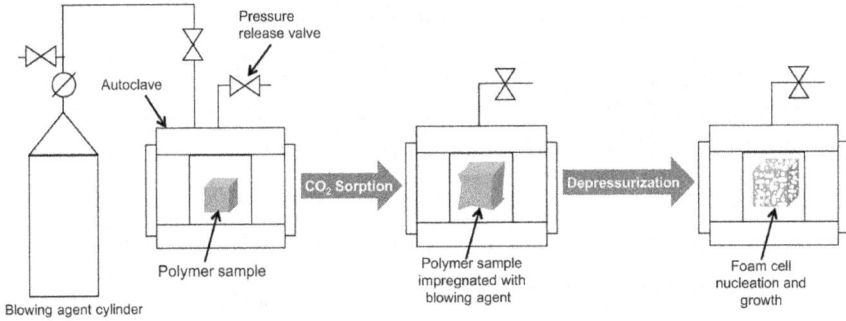

FIGURE 7.4   Pressure-induced batch foaming process (Okolieocha et al., 2015).

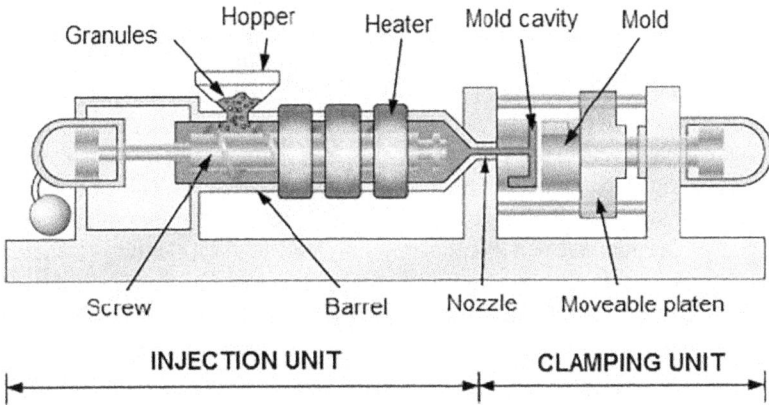

FIGURE 7.5   The injection moulding system (Svečko et al., 2013).

gas combination is injected into the mould cavity after the mould cavity has been compressed to roughly 6 MPa. After that, injection is used to fill the mould cavity. The gas is then released from the mould cavity and the mould is opened, causing foaming. Different degrees of mould opening resulted in foamed samples of varying thicknesses, and hence void percentages (Ameli et al., 2014). Less material is used, dimensional stability is improved, cycle time is reduced, and energy consumption is reduced using the foam injection moulding method. It may also help to enhance mechanical characteristics like fatigue life and impact resistance. For several different thermoplastics, extensive study has been done on FIM and characterization (Ameli et al., 2014).

Barrel temperature, mould temperature, screw speed, barrel pressure, metering time, injection flow velocity, the quantity of blowing, injection pressure, mould opening delay time, degree of mould opening, and gas counter-pressure are all significant factors in injection moulding. To successfully increase the homogeneity of the mixture to fulfil the requisite characteristics of the polymer nanocomposite foams, two preparation procedures were used.

1. The rate of concentration balance during the homogenization stage. At the polymer homogenization stage, the rate of concentration of the polymer/nanofillers and the blowing agent is influenced by the polymer matrix's temperature and chemical composition, and also, the blowing agent (Standau and Altstädt, 2019). Temperature affects the diffusion coefficient, which increases as the temperature rises.

2. The mass flow of molecules across a polymer matrix in a time-dependent manner. The mass flow of molecules across a polymer matrix in a time-dependent manner has been explained by Fick's equation, in which the diffusion rate is regulated by the concentration of the blowing agent (Sørensen and Spazzafumo, 2018, Standau and Altstädt, 2019). The saturation limit restricts the quantity of blowing agents that may be dissolved in the polymer (polymer dependent). When saturation rises, the local concentration changes, and therefore the diffusion rate decreases. As a result, the pressure and the gas concentration in the polymer are inextricably linked. The solubility coefficient could be represented as the pressure and temperature-dependent function at high pressures and temperatures.

Dispersed clay platelets in polymer nanocomposite foams not only change the physicochemical properties and the expansion behaviour of closed-cell foams but also create smaller and more isotropic cells by improving the thermal and mechanical properties of the foamed nanocomposite compared to the pristine polymer foam. Uniformly dispersed clay platelets enhance polymer melt strength, particularly during cell wall formation and stretching, and also stabilize the overall cell structure so that cell coalescence is inhibited. The nanofillers function as nucleating agents in the creation of bubble foams in the batch process using CO_2 as the physical foaming agent. In all cases, a small quantity of clay nanoparticles reduces the cell size of foams and at the same time increases the cell density (Sinha Ray, 2013, Wang et al., 2019).

The preparation procedure for flame retardant polystyrene foams has been described in which a co-rotating twin-screw extruder having a working temperature of 180°C at a rotation velocity of 150 rpm was used to premix polystyrene with nanofiller (flame retardants). In a stainless-steel autoclave, the dry extruded pellets were saturated with $ScCO_2$ for 4 h at 45°C and 12 MPa pressure. Hot-press moulding of saturated polystyrene granulates at corresponding foaming temperatures under 20 MPa was utilized to create foams. The flame-retardant foams were produced via fast pressure decrease after 15 min (Wang et al., 2019).

Nanoparticles are first evenly distributed in one or chemically combined monomers to create thermoset nanocomposite foams. Other monomers are added to the mixture to make it froth. Physical or chemical blowing agents can all be used to induce the foaming of polymers. Surface modification of nanoparticles, similar to the creation of thermoplastic nanocomposite foams, is necessary for successful nanoparticle distribution in a matrix. The majority of the studies employed functional surface modifiers on layered silicates, which can bond with one or more reactant molecules to generate an intermediate that promotes uniform nanoparticle dispersion in the polymer matrix in the course of foam formation (Lee et al., 2005). Compatibilizers are used to improve nanofiller dispersion in the polymer matrix while making certain nanocomposite foams. A compatibilizer and clay might broaden and narrow the cell-size distribution, respectively. The clay and the compatibilizer both decrease foam density while enhancing cell density. Because of the high degree of nanofiller dispersion, the polymer–nanofiller contact area is usually greater, resulting in increased cell nucleation. Furthermore, greater nanofiller delamination results in a higher melt viscosity, which limits gas flow in the cells, resulting in a larger number of cell sizes (Sinha Ray, 2013).

7.3   EFFECT OF NANOFILLERS ON POLYMER NANOCOMPOSITE FOAMS

The use of inorganic fillers in polymer matrixes to improve polymer characteristics has been widely employed (Han, 2015). Environmentally friendly gas like supercritical carbon dioxide ($ScCO_2$) is a good alternative to the conventional chlorofluorocarbon used as a blowing agent. However, the poor solubility and high diffusion rate of CO_2 in most polymeric materials make manipulation of foam cell shape/architecture more challenging. Because nanofillers produce additional nucleation sites in the polymer, a tiny quantity of well-dispersed nanofillers such as nano-clay, nanoparticles, and nanofibers in the polymer function as nucleation sites that aid the bubble nucleation procedure. As a result, the foam structure has a greater pore density and reduced pore size, allowing for a greater surface area (Lee et al., 2005, Sundarram et al., 2015). Gas diffusion in polymer matrices can be reduced by plate-like nanoparticles. Furthermore, the incorporation of nanoparticles in polymer foams may improve mechanical and physical characteristics, thermal degradation temperature, and fire resistance properties. Novel nanocomposite foams created by combining functional nanoparticles with the supercritical fluid foaming method result in a new class of multifunctional nanocomposite foam having lightweight and increased strength that can be applied widely in a lot of applications (Lee et al., 2005).

Graphene polymethylmethacrylate nanocomposite foam containing 1.8 vol% graphene sheets was produced and a microcellular regular cell size 5 µm was uniformly distributed throughout the foam matrix (Zhang et al., 2011). The cell size distribution of the almost spherical cells ranged from 1 to 10 µm. Compared to polymer foams with bigger cells packed with carbon nanofibers and nanotubes, this microcellular structure might improve the foam's electrical and mechanical characteristics. Furthermore, it has been found that the incorporation of graphene nanoplatelets transformed the cell shape from closed to open when the percentage of nanofiller was raised. Tobias and co-workers also discovered that adding a tiny quantity of nano-clay enhances the melt strength of the polymer nanocomposite foam, occasioning a substantial rise in foam cell density with an expansion ratio of 8. Due to increased melting temperature and the abundant cell nucleation sites, higher nano-clay concentration of up to 5% wt% leads to improved foaming ability and an expansion ratio of up to 20 with foam cell density up to 1089% cells/cm_3, resulting in a more even foam cell shapes (Yuceturk et al., 2018). The dispersion of nanofillers is critical for effective contact with the polymer. To guarantee the excellent dispersion of nanofillers in the polymer matrix, modifications to the polymer or nanofillers were made by decorating the surface of the polymer or nanofillers. In particular, using solvent-free nanofluids to address aggregation and optimize the outstanding characteristics of nanoparticles in composites has been proposed as a viable solution (Gao et al., 2020). Ball-milling, shear-mixing techniques, melt mixing, ultra-sonication, and other effective techniques have been developed (Tan et al., 2020, Xie et al., 2021). The use of solvent-free nanofluids, in particular, has been touted as an effective approach to combat aggregation and optimize the outstanding characteristics of nanoparticles in composites (Yao et al., 2018). Gao et al. (2020), multifunctional nanocomposite foams by the surface modification of polyether amine and organosilane chains to create well-dispersed functionalized multi-wall carbon nanotubes (MWCNTs) that were then employed as a filler in epoxy/f-MWCNTs microcellular foam preparation.

7.3.1 PARAMETERS THAT INFLUENCE FOAM CELL STRUCTURE

Temperature and pressure during foaming have an important influence on the structure and growth of foam cells (Standau and Altstädt, 2019). Because the solubility and diffusivity of blowing agents into polymer matrices are temperature and pressure-sensitive, these factors can influence to a greater extent the content of blowing agents that can dissolve and diffuse into the foam matrix within a certain period. The consequence of pressure and variable temperature on the solvability of blowing agents differ according to the kind of agent involved. For example, when pressure rises, the solubility of CO_2 and N_2 in a polymer nanocomposite matrix also increases. However, when the temperature rises, the solubility of CO_2 in the sample drops, but the solubility of N_2 in the sample increases. It is, therefore, important to note that temperature and pressure alter the foam's crystal structure, which has an impact on the cell structure. Most polymers are semi-crystalline, and therefore, blowing agents are dissolved only in the amorphous regions, leaving the crystalline portion unaffected. It has been observed that when pressure rises, inter-lamellar areas expand due to the blowing agent, allowing more room for cell development and resulting in foams having a more uniform shape in the amorphous and crystalline regions (Standau and Altstädt, 2019). The separate impacts of temperature and pressure are examined in addition to their combined effects.

The temperature has a momentous influence on the foaming outcome. The high rigidity of the nanocomposite matrices makes it almost impossible to expand at low temperatures. With a temperature rise, the melt strength of the matrix falls, and the deformability of the matrix rises, allowing cells to be enlarged more easily by the blowing agent. At a higher temperature, the blowing agent diffuses more easily in the matrix, which promotes cell development. Generally, blowing agents diffuse out of matrices and into the environment when the processing temperature is extremely high instead of contributing to foam cell formation and growth. As a result, an optimal temperature is required for the nanocomposite foam to remain adequately soft to be able to expand while retaining a suitable quantity of blowing agents (Standau and Altstädt, 2019).

Foaming is also affected by saturation pressure and pressure decrease rate. Besides the effect on the solubility of the blowing agent, the foaming rate and pressure also affect the energy barrier. It is important to note that Gibb's energy barrier to nucleation is lower when the foaming pressure is extremely high, hence, the pressure must be reduced considerably to approximately that of an atmospheric pressure to facilitate cell nucleation (Standau and Altstädt, 2019).

The cell structure morphology and density of tungsten/polymethyl-methacrylate (W/PMMA) composite microcellular foams with various W loadings were investigated using the constitutive model. The microcellular composite foams were prepared using melt blending and scCO$_2$, as shown in Figure 7.6. The W/PMMA composite microcellular foams showed substantially smaller cell widths than pure PMMA microcellular foams, while the former had a much greater cell density (Zhu et al., 2019). Cell nucleation, according to conventional nucleation theory, requires overcoming a high nucleation energy barrier; nevertheless, cells do not nucleate easily in pristine PMMA foam. As a result, when there is insufficient gas in the PMMA matrix, no new cells are formed; instead, the gas enters the existing cells resulting in dramatic growth in cell sizes. When W particles are introduced, the nucleation energy barrier is reduced considerably, and more cells are formed. Following cell stabilization, the unused residual gas in the PMMA matrix is dispersed into the already-created cells, resulting in cell sizes that are lower than those achieved with pure PMMA. Therefore, W particles can function as heterogeneous nucleation agents, lowering the energy barrier and resulting in greater cell densities (Zhu et al., 2019).

Similarly, the impact of pressure on the produced foams' cellular architecture and hardness was investigated and the results indicate that a rise in saturation pressure results in a decrease in cell diameter and, as a result, a rise in cell density. The cell nucleation and growth rate explain the fluctuation in cell diameter about the

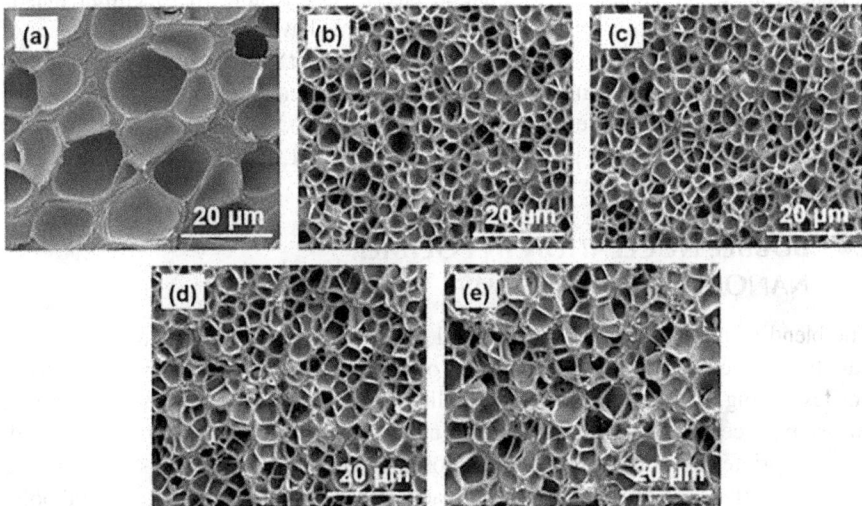

FIGURE 7.6 FESEM micrographs of the cell morphology of W/PMMA composite microcellular foams having varying contents of W manufactured at a saturation pressure of 18 MPa, with a foaming temperature of 80°C at 10 s: (a) pure PMMA foam, (b) 10 wt% W/PMMA foam, (c) 20 wt% W/PMMA foam, (d) 40 wt% W/PMMA foam, and (e) 60 wt% W/PMMA foam (Zhu et al., 2019).

saturation pressure. When the saturation pressure increases, the solubility of CO_2 in most polymer matrices also increases, which most often results in the creation of additional nucleation sites when the pressure was released (Arab-Baraghi et al., 2014). Usually, the initial number of nucleated bubbles controls the amount of CO_2 molecules allocated per bubble. When the foaming mixture gets saturated at higher pressure, more nucleation sites become readily available, and therefore fewer CO_2 molecules are needed to create the bubbles throughout the growth phase, which mostly results in smaller and more uniform cells. The foaming temperature and pressure determine the saturation time (thus, the time taken for the blowing agent to dissolve and diffuse into the polymer/nanofiller combination to achieve equilibrium). When the distribution of blowing agents inside the nanocomposite is not complete or uniform, poor foam expansion and nonhomogeneous cell structure occur fast. As a result, more blowing agent absorption is necessary to ensure optimal cell nucleation and development in the depressurization phase.

Hence, the symmetry variations, the packing density, and space-filling are usually both quite high. However, when the actual systems' shape isn't spherical, but rather hexagonal closest pack (hcp), this can take up more than 95% of the volume. The foam properties are influenced by cell kind, as well as cell size, thickness, and density (Rende et al., 2013). To offer good mechanical properties, such as rigidity and strong thermal insulation, foam cell width ought to be decreased while maintaining a uniform cell shape. The thermal insulating qualities improve as the average cell diameter decreases. The rigidity of the foam, and hence its mechanics, is not determined solely by the cell diameter, but also by the framework thickness, which may be changed by reducing the density (higher density results in higher foam stiffness) (Chen et al., 2015, Standau and Altstädt, 2019). As a result, microcellular plastic foams are the subject of a lot of studies. These foams have a regular cell diameter of approximately 1–10 m, a high cell density, and a weight reduction of up to approximately 5–95%. It is important to state that closed or open-cell configurations can be purposefully created by selecting an appropriate method and altering the chemical composition of the polymer phase (Standau and Altstädt, 2019).

7.4 BUBBLE NUCLEATION IN POLYMER NANOCOMPOSITE FOAMS

The blend of nanofillers and supercritical fluid or chemical-blowing-agent foamers has the prospect to produce a new class of lightweight, robust, and versatile composites having wide applications. The bulk of nanocomposite foams is created in a two-step process. The process starts with nanocomposite synthesis and finishes with an appropriate foaming technique. For commercial operations, the use of foaming agents directly is the most common approach. When gas is injected into a solid polymer and subsequently depressurized, extrusion, injection moulding, and compression moulding can create foams in the liquid–melt state. For most foams, various optimal cell morphologies (e.g., tiny vs. huge cells, open vs. closed cells) ought to be obtained by correct bubble nucleation and growth management (Feldman, 2010, Sinha Ray, 2013).

7.4.1 Blowing Agents

A critical component of foaming is the blowing agent. Physical and chemical foaming agents are commonly utilized. Compounds that gasify under foaming conditions are referred to as physical foaming agents, whereas chemical foaming agents are often reactive species that generate gases during the foaming process (Lee et al., 2005). The majority of chemical foaming agents are complexes that release gases due to the chemical reaction that goes on (thermal decomposition by the exothermic reaction, mostly by the release of CO_2, CO, N_2, NH_3, sodium bicarbonate, citric acid and derivatives, azodicarbonamide, etc.) (Standau and Altstädt, 2019). Chemical blowing agents have lower expansion ratios than physical propellants (process and application dependant), and a solid residue lingers in the polymer. Conventionally, the following mechanisms are identified:

a. An irreversible reaction resulting in the emission of gas. AB → C + Gas. For example, azodicarbonamide ⇒ CO_2, CO, N_2, NH_3
b. An equilibrium reaction causing the release of gas: AB ↔ C + Gas. For example, alkaline or alkaline earth metal oxides and bicarbonate ions that may result in CO_2
c. A mixture of various constituents that emit gas as a result of a chemical reaction. Thus A + BG → AB + Gas. For example, combined sodium bicarbonate and citric acid may produce CO_2 and H_2O (Standau and Altstädt, 2019)

Volatile compounds like chlorofluorocarbons, volatile hydrocarbons, and alcohols, and inert gases like nitrogen, carbon dioxide, argon, n-Butane, and water are all physical foaming agents. They produce foam in materials through physical processes by changing their state, for instance, from liquid to gas, and by evaporation, or through desorption at high temperatures or lower pressure (Arab-Baraghi et al., 2014, Fan et al., 2019, Kuang et al., 2016, Wang et al., 2019). Extrusion, injection moulding, and compression moulding can generate liquid/melt foams, or gas can be pushed into a crystalline solid polymer after which it is depressurized to form solid foam. Physical or chemical blowing agents can be used in both procedures. Propellers must generally meet specific criteria, such as excellent solubility, odour neutrality, environmental friendliness, and economic viability. Nowadays, physical foaming techniques with hydrocarbons or chlorofluorocarbons as blowing agents are commonly used to make polystyrene foams. The replacement of traditional foaming agents, however, is a key technical breakthrough owing to environmental and safety issues. This is also why chlorofluorocarbons have been phased out of usage. Even while chlorofluorocarbons are effective propellants, they deplete the ozone layer catalytically and have a high carbon footprint. Only a few propellants are suitable for industrial use based on the requirements. Both volatile solvents, such as ketones or hydrocarbons, and gases can be employed to foam polymers. The gases are inert chemicals like N_2, CO_2, or helium that are dissolved in the melted polymer at high pressure. This reduces the solubility of the blowing agent substantially when the pressure drops, causing the polymer to melt to foam physically (mechanical foaming) or by gas expansion. This process is flame resistant, inexpensive, harmless, and ecologically benign. CO_2 is a typical blowing agent suitable for the manufacture of polymeric foams. The advantage of CO_2 for the manufacture of polymeric foams is that the critical state (temperature) is moderate and simple to attain (thus the critical temperature is around 31.1°C at a critical pressure of 7.38 MPa. Most materials absorb supercritical carbon dioxide ($ScCO_2$) to saturation and then undergo rapid depressurization at a fixed temperature to produce polymeric foams. Some studies have investigated the use of $ScCO_2$ for the fabrication fire resistant polymeric foams. Such polymers include poly(lactic acid), polypropylene, and polyethylene (styrene-co-acrylonitrile). $ScCO_2$ foaming technique was used to fabricate poly(styrene-co-acrylonitrile) foam modified by clays/melamine polyphosphate. A substantial decrease in heat release rate was achieved due to the synergistic impact of the clays/melamine polyphosphate (Urbanczyk et al., 2010). It is important to state that the type of blowing agent used in the manufacture of foams usually has an impact on the characteristics and the uniformity of foam cells. Porous polymeric materials can be fabricated into foams using blowing agents and other techniques such as phase inversion, leaching, and thermal breakdown. The majority of these technologies are only appropriate for thin-film product preparation (Lee et al., 2005).

Thermoplastic nanocomposite foams are generally made with chemical blowing agents. The finished product must have a closed-cell configuration with thin cells, regardless of the foaming technique utilized. To create a closed-cell configuration, the cell development ought to be carefully monitored by managing the degradation temperature of the chemical foaming agent as well as careful control of the melt viscosity of the polymer. When the temperature is too high foaming agent decomposes fast and the melt strength of the polymer decreases dramatically, resulting in flocculation and cell rupture. When the temperature is too low, the rate of degradation of the foaming agent slows down, which necessitates a longer foaming period, and the melt viscosity and strength of the polymer drop, occasioning cell amalgamation and subsequent rupture (Sinha Ray, 2013).

7.5   PROPERTIES OF POLYMER NANOCOMPOSITE FOAMS

7.5.1   MECHANICAL PROPERTIES OF POLYMER NANOCOMPOSITE FOAMS

Mechanical characteristics of a polymer foam, such as tensile strength and storage modulus, are enhanced following the addition of nanofillers, as seen in solid polymer nanocomposites. They expand as the amount of nanofillers increases. In the case of foams, Antunes et al. (2010) found that the storage modulus rose very marginally increasing carbon nanofiber loading for equal relative densities. Nevertheless, when more nanofibers were added, the specific storage modulus, or storage modulus relative to the density of the foam, rose significantly, demonstrating the carbon nanofibers' effectiveness as mechanical reinforcements. Apart from carbon nanofillers, a notable reinforcing impact of montmorillonite particles in polymer nanocomposite foam with a more than 30% increase in storage modulus was observed (Madaleno et al., 2013). In comparison to the equivalent basic PA-6 microcellular sample, nylon 6/clay nanocomposite microcellular foams had greater tensile strength. Contrarily to neat PVC foam, the PVC/3% cloisite 30B nanocomposite foam blown by three parts by weight of CO_2 exhibits a 17.9% improvement in tensile strength, a 25.9% increase in bending strength, and a 250% increase in elongation ratio (Lee et al., 2005). Extruded Polystyrene/carbon nanotube nanocomposite foams' tensile characteristics were also studied. In comparison to bulk polystyrene, polymer foams have an analogous foam density of 0.6–0.7 g/cm³. The tensile modulus of plain polystyrene foam drops by 40% (1.26–0.74 GPa) when the weight is reduced by 37%. Despite this, the increases in the tensile modulus by 28% (0.74–0.94 GPa) in the presence of 1 wt% carbon nanofiller. When the fibre concentration is raised to 5% by weight, the tensile modulus improves to 1.07 GPa, which is equivalent to the bulk polystyrene modulus (1.26 GPa). The reduced modulus was used to relate the two samples to equalize the influence of foam density on mechanical characteristics. The decreased modulus of polystyrene/carbon nanotubes foams is significantly higher than the bulk polymer composite because of the lower densities.

7.5.2   FLAME RETARDANCY OF POLYMER NANOCOMPOSITE FOAMS

To reduce the amount of fire damage to polymer composites, there has been a lot of interest in creating fire-resistant polymers. Designing and synthesizing thermally stable polymers that are less prone to break down into flammable gaseous species when subjected to high temperature is critical from this perspective. Polymer nanocomposite foams are polymers containing nanometre-sized particles finely distributed throughout the polymer matrix. Nano clays such as layered silicates, montmorillonite, and inorganic nanoparticles such as metal oxides, metal hydroxides/hydrates, and metal carbides/nitride, as well as graphene, carbon nanotubes, and inorganic nanoparticles such as metal oxides, metal hydroxides/hydrates, and metal carbides/nitride, have been commonly used as fillers to reduce the susceptibility of foams to fire (Ahmadzadeh et al., 2016, Bagaria et al., 2019, Han et al., 2016, Kim et al., 2020, Urbanczyk et al., 2010, Yao et al., 2018). The fillers lower the mass loss rate of the polymer by establishing a protective barrier inside the polymer matrix (Kim et al., 2021). Because of their flame retardant and thermal stability, solid polymer nanocomposites, particularly those containing flame retardant nanofillers, have a wide range of applications. This characteristic is mostly connected to the fillers in composites that increase the polymer's retardancy. The contribution of each type of nanoparticle to the polymer's fire resistance varies and is solely dependent on the chemical structure and shape of the nanoparticle (Kim et al., 2021). The influence of expandable graphite (EG) and melamine phosphate (MP) on the decomposition behaviour and fire performance of polymer foams was studied and the flame retardant was found to produce inert gases and accelerate the development of char from nanocomposite in polymer nanocomposite foams. The thick char layer function as a physical barrier, which reduces the heat release of polymer nanocomposite foams. Additionally, the incorporation of triphenyl phosphate (TPP) and hexaphenoxycyclotriphosphazene (HPCTP) enhances the foaming and fire inhibition properties of composite foam. The flame-retarded foams (with 25 wt.% MP/EG) achieved HF1 and a V-0 rating. A limiting oxygen index (LOI) of 30.1 and 29.6% were achieved for the TPP and HPCTP composite foams, respectively. To increase gas-phase flame suppression, TPP and HPCTP can create phosphorous oxides and transient phenoxyl radicals to increase the flame retardant effect in both condensed and gas phases. The microcalorimeter test gives a quantitative evaluation of the synergistic action of TPP and HPCTP and further improves the flame resistance of polystyrene foams. Further study using X-ray photoelectron spectroscopy (XPS) on char residues from polystyrene foams reveals the occurrence of P–O–C bonds and other thermally stable macromolecular structures (Wang et al., 2019).

The presence of MP/EG flame-retardant can increase the onset degradation temperature of polystyrene foams in nitrogen, implying improved thermo-oxidative stability. The thermo-oxidative stability of foams decreases gradually as the proportion of MP/EG in the composite foam increases due to lower start decomposition temperatures of EG (190°C) and MP (285°C), respectively. A redox reaction between H_2SO_4 and EG creates a vast number of blowing gases, resulting in a vermiculate EG structure having enormous volumes (CO_2, SO_2, and H_2O). When a little quantity of HPCTP is added, the char residue increases, promoting the development of a thick barrier that strengthens the flame-retardant action in the condensed phase. A char coating increases the escape time of volatile products increases, causing the thermal oxidation of polystyrene to be accelerated. When it comes to flame-retardant foams, the peak of the heat release rate and the total heat release (THR) values drop as the percentage of flame-retardants increases. Adding a little TPP or HPCTP to the polymer foam composites decreases the peak heat release rate (PHRR) and THR values even further. The PHRR and the THR values of HPCTP foams are 496 W/g and 27.9 kJ/g, respectively, which are 46.6 and 26.2% lower than those of plain polystyrene foam (Wang et al., 2019). The cohesiveness of the protective carbonaceous layer generated at the onset of combustion might also be linked to the reduction in the PHRR and the fire suppression ability of the flame retardant (Urbanczyk et al., 2010). Although the foams are made up of 90% voids, clay and melamine polyphosphate enhance the creation of a solid protective char layer capable of acting effectively to shield the unburned foam layer from further thermal degradation.

7.6 ELECTRICAL PROPERTIES OF POLYMER NANOCOMPOSITE FOAMS

The bulk of electrically conductive nanocomposite foams is made out of thermoplastic matrices with a glass transition temperature between 65 and 105°C and melting temperatures between 65 and 240°C. PVF, PMMA, PS, and low-density PE are among the thermoplastic matrices employed. The porosity of these foams' ranges from 15 to 85%, while the pore size ranges from a few hundred microns to a few microns. Depending on the kind and number of nanoparticles employed, the electrical conductivity varies considerably between investigations, ranging from 0.1 to 6 10 8 S/cm. Generally, a higher conductivity is associated with a higher loading of the conductive phase. Despite having strong electrical conductivities for electromagnetic interference shielding applications, the majority of such nanocomposite foams have low utility temperatures (Sundarram et al., 2015).

Polyethylenimine/multi-walled carbon nanotube (PEI/MWCNT) nanocomposite foams were produced and the electrical and mechanical properties were studied.

The nanocomposites were prepared with dichloromethane (DCM) as the solvent in a solution-based approach and foamed with CO_2 as the blowing agent. The impact of various foaming conditions and residual DCM solvent was investigated (Kim and Li, 2013). The mechanical characteristics of nanocomposites were shown to be considerably influenced by residual solvent. The relative density of the composite foams was observed to affect their electrical conductivity. Electrical conductivity was greater in foams with a higher relative density. The glass transition temperature of the produced PEI nanocomposite foam was similar to that of the plain PEI. With 2 wt.% MWCNTs loadings, the PEI composite foam retained a high electrical conductivity of up to 10 7 S/cm³ despite the volumetric expansion resulting from the foaming process. Even with a low relative density of 45%, such foams are acceptable for electrostatic dissipative purposes (Sundarram et al., 2015).

An asymmetric shielding network with electrical and magnetic gradients might be established in conductive polymer nanocomposite foams, and in such a scenario, the release of electromagnetic waves might be further controlled by a combination of multi-interface absorption and hysteresis loss absorption, reducing the R-value. The standard template or pre-constructed conductive network approach, on the other hand, makes it impossible to build a multi-layered conductive network with electromagnetic gradients (Duan et al., 2020). Nanocomposite foam with nano-additives in nanocomposite foam has a greater electrical conductivity than their solid counterparts, according to Antunes *et al*, indicating that foaming might be a valuable technique in producing an electrically conductive network (Antunes et al., 2010).

7.7 APPLICATION OF POLYMER NANOCOMPOSITES FOAM IN ACOUSTICS

Manufacturing civilization, civil infrastructure, and transportation have all resulted in substantial noise pollution, which is regarded to be hazardous to human health and the environment (Han, 2015, Lercher, 2019). Noise pollution is becoming a more serious worldwide health issue. Considering the many adverse effects of noise pollution, several kinds of research geared towards the avoidance and control of noise have been conducted worldwide (Geravandi et al., 2015, Tripathy, 2008, Wang et al., 2005). To address this issue, materials with great sound absorption capacity are being developed. Because of their viscoelasticity properties, lightweight, and formability, polymer materials like polyurethane have indeed been widely utilized for sound insulation and absorption (Cao et al., 2018, Fu et al., 2021). Nonetheless, because of its lightweight and low elastic modulus, the soundproofing characteristics of pure polymer are still insufficient to fulfil the stiffness and mass laws of industrial applications. Many researchers have been interested in polymer modifications for excellent noise reduction capabilities (Han, 2015). Studies have shown that non-acoustic characteristics like porosity, density, and air-flow resistance may be used to understand the acoustic properties of foams (Zhu et al., 2018). Porous acoustical materials, such as polymer nanocomposite foams have channels, fractures, or cavities that enable sound waves to pass through. Thermal loss, produced by air molecules rubbing against pore walls, and vicious loss, induced by the viciousness of airflow inside the materials, both contribute to the dissipation of sound energy (Cao et al.,

2018). Foams' interconnected open pores play an important role in determining not only the mechanical but also acoustic properties. There is a link between foam cell porosity and sound absorption capabilities (Hyuk Park et al., 2017, Park et al., 2017).

In terms of sound absorption, foam with a larger cell porosity performs best. Porous materials have a high absorption coefficient due to their irregular pore morphology. Pure polymer foams, on the other hand, have poor low-frequency sound absorption due to their distinct pore geometries (Ji et al., 2019). Functional materials like nanofillers, granular microspheres, and crumb rubber particles can be added to improve the shortcoming of pure polymer foams. The most common method for noise reduction is to use sound-insulating or absorption materials. Because sound insulation materials can only modify the sound propagation path, but sound absorption materials may directly absorb sound waves, various studies have concentrated on improving the sound absorption capabilities of polymer foams (Lee et al., 2012).

Compared to metal-based materials, conductive polymer nanocomposite foams are preferable, especially for lightweight purposes such as aircraft, spacecraft, and cars. The porous nature of foams provides additional benefits for EMI shielding due to improved magnetic energy absorption via wave scattering (Sundarram et al., 2015). The isocyanate index, cell size, density, and molecular weight of polyols were used to investigate the sound absorption ratio of polyurethane/nano-silica foams. Across the entire frequency range, it was noticed that the sound absorption ratio of polyurethane/nano-silica foams improves as the nano-silica concentration of the foam increases. As the isocyanate index, cell size, and density all dropped and the sound absorption ratio of polyurethane/nano-silica foams increased. Because the molecular weight of the polyol is reduced, the glass transition temperature of polyurethane/nano-silica foams is centred around ambient temperature, resulting in good sound absorption (Lee et al., 2012).

Monomers such as ethylene propylene diene of various contents and hardness are also employed as fillers to create foams with a variety of pore topologies and sound absorption characteristics. The addition of monomeric particles to foam results in smaller holes, higher density, and greater air-flow resistance, according to the findings. Simultaneously, the foam composites have improved sound absorption characteristics in the middle-frequency range, and better values may be achieved at lower frequencies as the monomeric particle concentration increases. Sound absorption characteristics are significantly influenced by the toughness of the monomer particles, notably in the medium frequency range. This indicates that the morphologies of foam pore morphogenesis affect sound absorption characteristics (Zhu et al., 2018). Plate-like nanofillers, like bentonite, organophilic clays, and sodium montmorillonite intercalated with poly (ethylene glycol), might be useful for fabricating nanocomposites meant to improve the sound-dampening effect of flexible polyurethane foam, especially when it has an open-cell structure. Within the high-frequency region, the plate-like fillers significantly enhanced sound damping (Sung et al., 2007). Apart from the increased interfacial adhesion and stiffness of the polymer nanocomposite foam due to the addition of nanofillers, one of the main reasons that improve the sound insulation performance is the increased interfacial adhesion of the polymer nanocomposite foam (Kim et al., 2013, Lee et al., 2009).

7.8 CHARACTERIZATION OF ACOUSTIC POLYMER NANOCOMPOSITES FOAM

The basic approach for evaluating soundproofing equipment is sound transmission loss (STL) calculated as per equation (7.1),

$$STL\,(dB) = 10\log_{10}\frac{E_i}{E_t} \tag{7.1}$$

The efficiency of the specimen's STL is calculated by multiplying Ei by Et, where Ei is the incident sound energy and Et is the transmitted sound energy.

The test is based on ISO 10534-2 (1998) and ASTM E1050-12 (2012), which uses the transfer function technique. The STL refers to the variance between the incident and transmitted sound power levels. As illustrated in Figure 7.7, the STL configuration includes many microphone impedance tubes. It has a 30-mm-diameter tube (small tube) for high-frequency testing (1600–6400 Hz) and a 60-mm diameter tube (large tube) for low and medium-frequency testing (100–2500 Hz). The average of three or more samples is generally provided, along with error bars (Asadi Khanouki and Ohadi, 2018, Han, 2015).

When sound hits a wall, it vibrates in reaction to changes in air pressure. This vibration energy dissipates and rises according to the wall's weight during the transmissible process from the inside to the outside of the wall. This connection has been dubbed the Mass Law of Sound Insulation (Lee et al., 2009).

The transmission loss (TL) of sound-insulating material is determined using the mass Law in equation (7.2).

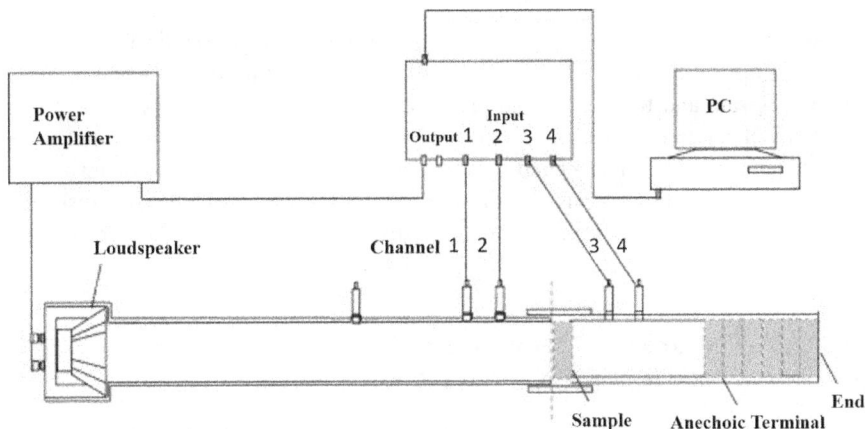

FIGURE 7.7 Schematic of the four-microphone impedance tube used to measure sound transmission loss (Han, 2015).

$$TL(\theta, f) = 10 \log_{10} \left\{ 1 + \left(\frac{\omega m \cos \theta}{2\rho c} \right)^2 \right\} \qquad (7.2)$$

by which ω, m, and θ are the angular frequency, the surface mass of the specimen's unit area, and the angle of incidence, respectively.

The normalized masses are computed using the predetermined densities and volumes of the manufactured specimens, and the densities of the produced specimen are derived using the measured masses and volumes of the produced specimen. Additionally, the stiffness of the polymer nanocomposite foam has an impact on sound insulation effectiveness. As a result, a micro indenter is used to test the stiffness of the specimens under ISO 14577-1 (Lee et al., 2009). By examining pictures captured by a scanning electron microscope, the microstructural features of foam samples obtained from the foam structure are retrieved (scanning electron microscope (SEM)). For imaging, a few tiny specimens with an average size of 5 × 5 × 2 mm are produced and gold coated. The microstructural parameters of cell size (Cs), strut thickness (t), strut length (l), and reticulation rate (r) are examined using a SEM and image analysis software. Samples of polymer nanocomposite foams might be cryogenically broken in liquid nitrogen and studied with a scanning electron microscope to explore the cell structure of foams (Asadi Khanouki and Ohadi, 2018, Lee et al., 2009).

Five non-acoustic characteristics are required for modelling sound transmission in porous materials: porosity (ϕ), static airflow resistivity (σ), static tortuosity ($\alpha\infty$), viscose and thermal characteristic lengths (Λ, Λ'). These non-acoustic characteristics have received little attention in the research of nanocomposite foams, although they are essential for showing sound absorption behaviour in pristine as well as nanocomposite foams. Porosity refers to the amount of pore space occupied by a fluid phase in a porous structure such that if VT is the total volume of a porous material and Vf is the volume of the fluid phase, then porosity = Vf/VT may be used to calculate it (Asadi Khanouki and Ohadi, 2018)

The resistance of a porous substance to a fluid flow is known as static airflow resistivity. This amount is acquired through direct measurement for the sake of precision. The airflow rate and the differential pressure between two sides of cylindrical test specimens of 3-cm diameter and 3-cm thickness are measured using a permeability metre device (Asadi Khanouki and Ohadi, 2018). Other methods of characterization include X-ray diffraction, transmission electron microscopy, and Fourier transforms infrared spectroscopy (Ahmadzadeh et al., 2016) have been used to study the crystal structure and morphological characteristics of polymer nanocomposite foam.

7.9 CHALLENGES IN POLYMER NANOCOMPOSITE FOAM APPLICATIONS IN ACOUSTICS

Some specific challenges of nanocomposite applications in acoustics are as follows:

1. Good dispersion and the compatibility of nanofillers with the polymer matrix: Nanofillers tend to clump together if they aren't equally distributed. The characteristics of the polymer nanocomposite foam are impacted as a

result. To achieve functional polymer nanocomposite foam, good interaction between the nanofiller and the matrix is essential. Sometimes, additives are used to improve dispersion which may /may not change the polymer nano-composite foams' anticipated characteristics. To help expand the useful-ness of polymer nanocomposite foams in different industrial applications, research into strategies for enhancing the polymer-nanofiller interaction and the effect of additional chemicals or methods for improving disper-sion and compatibility is greatly needed, just as it is for normal polymer nanocomposites.

2. Insufficient loading of conductive materials in polymer foams: Low loading of conductive nanofillers in nanocomposite foams intended for electrical applications might result in linear behaviour with a frequency characteristic of insulating materials such as plain polymers. Because the nanofillers are too far apart to enable electrical conduction, this implies that the matrix controls the composite's electrical characteristics (Antunes et al., 2010).

3. Degradation during use. In addition, some of the polymers used to make nanocomposites are not easily degradable and might last for years in an anaerobic environment. Although it is unquestionably undesirable for com-posites to deteriorate during the application, their non-degradability poses a significant environmental risk after they have been used. Unlike thermo-plastic polymer nanocomposite foams, which may be recycled, thermoset nanocomposites cannot be reused. As a result, the search for more accept-able ways to dispose of old polymer nanocomposite foams is currently ongoing.

4. Inadequate interconnected pores via $scCO_2$ foaming. Salt leaching, ultra-sonic post-treatment, co-solvent, and processing parameter management have all been used in combination with gas foaming technology to improve pore connection, but further research and process optimization are still needed to regulate the scaffold architecture and characteristics. When $ScCO_2$ is used in a traditional microcellular foaming method, homoge-nous nucleation occurs initially when the pressure drops following $ScCO_2$ dissolution (Ahmadzadeh et al., 2016). During the cell development stage, the CO_2 is then consumed by nucleation sites, which increases the cell size. The limited impact of the crosslinking network on the mobility of molecu-lar chains hampers the growth of bubbles in an epoxy-based thermosetting system. As a result, making epoxy microcellular foam is rather difficult (Gao et al., 2020).

7.10   CONCLUSION

Polymer nanocomposite foams are functional products with wide applications due to their lightweight, electrical properties, insulation, mechanical properties, etc. The properties are related to many different factors such as the type of nanofillers incor-porated in the polymer matrix. The main distinguishing feature in the structure of polymer nanocomposite foam compared to solid polymer nanocomposite is its cells.

Two main types of cell structures in foam, some preparation methods, the effect of fillers, and the factors that affect the foam formation in nanocomposite foams have also been discussed. The closed and open cells of the foams are in different sizes, related to the type of blowing agent and other parameters such as temperature, pressure, and saturation time. The addition of nanofillers has not only improved the foaming process of polymer nanocomposite foam but has also expanded its application in acoustics. However, a higher amount of nanofillers tend to either negatively affect the properties or present no significant improvement. Nanofillers loading increases cell density and reduces cell size to mitigate the high nucleation efficiency during fabrication. The narrow operation window and the undesirable cell morphology, which limit neat polymer foams are mitigated by the incorporation of nanofillers. Among others, useful properties such as stiffness have been enhanced, which makes their application for sound insulation more convenient, especially in a high-frequency range. Due to the useful properties of polymer nanocomposite foams, they remain highly promising materials for applications in many areas such as the protection of human health and energy transmission/storage. With the advances in science and technology and the current research interest in nanocomposite foams, the possibilities of nanocomposite foam applications remain limitless.

REFERENCES

Ahmadzadeh, S., Keramat, J., Nasirpour, A., Hamdami, N., Behzad, T., Aranda, L., Vilasi, M. & Desobry, S. 2016. Structural and mechanical properties of clay nanocomposite foams based on cellulose for the food-packaging industry. *Journal of Applied Polymer Science,* 133, 2.

Ameli, A., Jahani, D., Nofar, M., Jung, P. & Park, C. 2014. Development of high void fraction polylactide composite foams using injection molding: Mechanical and thermal insulation properties. *Composites Science and Technology,* 90, 88–95.

Antunes, M., Realinho, V. & Velasco, J. I. 2010. Foaming behaviour, structure, and properties of polypropylene nanocomposites foams. *Journal of Nanomaterials,* 2010, 306384.

Arab-Baraghi, M., Mohammadizadeh, M. & Jahanmardi, R. 2014. A simple method for preparation of polymer microcellular foams by in situ generation of supercritical carbon dioxide from dry ice. *Iranian Polymer Journal,* 23, 427–435.

Asadi Khanouki, M. & Ohadi, A. 2018. Improved acoustic damping in polyurethane foams by the inclusion of silicon dioxide nanoparticles. *Advances in Polymer Technology,* 37, 2799–2810.

Bagaria, P., Prasad, S., Sun, J., Bellair, R. & Mashuga, C. 2019. Effect of particle morphology on dust minimum ignition energy. *Powder Technology,* 355, 1–6.

Cao, L., Fu, Q., Si, Y., Ding, B. & Yu, J. 2018. Porous materials for sound absorption. *Composites Communications,* 10, 25–35.

Chen, Y., Das, R. & Battley, M. 2015. Effects of cell size and cell wall thickness variations on the stiffness of closed-cell foams. *International Journal of Solids and Structures,* 52, 150–164.

Das, R., Pattanayak, A. J. & Swain, S. K. 2018. 7- Polymer nanocomposites for sensor devices. *In:* Jawaid, M. & Khan, M. M. (eds.) *Polymer-based Nanocomposites for Energy and Environmental Applications.* Woodhead Publishing.

Duan, H., Zhu, H., Gao, J., Yan, D.-X., Dai, K., Yang, Y., Zhao, G., Liu, Y. & Li, Z.-M. 2020. Asymmetric conductive polymer composite foam for absorption dominated ultra-efficient electromagnetic interference shielding with extremely low reflection characteristics. *Journal of Materials Chemistry A,* 8, 9146–9159.

Fan, X., Zhang, G., Li, J., Shang, Z., Zhang, H., Gao, Q., Qin, J. & Shi, X. 2019. Study on foamability and electromagnetic interference shielding effectiveness of supercritical CO2 foaming epoxy/rubber/MWCNTs composite. *Composites Part A: Applied Science and Manufacturing*, 121, 64–73.

Feldman, D. 2010. 10-Polymeric foam materials for insulation in buildings. *In:* Hall, M. R. (ed.) *Materials for Energy Efficiency and Thermal Comfort in Buildings*. Woodhead Publishing.

Fu, Y., Kabir, I. I., Yeoh, G. H. & Peng, Z. 2021. A review on polymer-based materials for underwater sound absorption. *Polymer Testing*, 96, 107115.

Gao, Q., Zhang, G., Fan, X., Zhang, H., Zhang, Y., Huang, F., Xiao, R., Shi, X. & Qin, J. 2020. Enhancements of foamability, electromagnetic interference shielding and mechanical property of epoxy microcellular composite foam with well-dispersed f-MWCNTs. *Composites Part A: Applied Science and Manufacturing*, 138, 106060.

Geravandi, S., Takdastan, A., Zallaghi, E., Vousoghi Niri, M., Mohammadi, M. J., Saki, H. & Naiemabadi, A. 2015. Noise pollution and health effects. *Jundishapur Journal of Health Sciences*, 7, 1.

Han, L., Wu, W., Qu, H., Han, X., Wang, A., Jiao, Y. & Xu, J. 2016. Metallic ferrites as flame retardants and smoke suppressants in flexible poly(vinyl chloride). *Journal of Thermal Analysis and Calorimetry*, 123, 293–300.

Han, T. 2015. Light-weight poly(vinyl chloride)-based soundproofing composites with foam/film alternating multilayered structure Part A Applied science and manufacturing. *Composites*, 78, 27–34–2015 v.78.

Hyuk Park, J., Suh Minn, K., Rae Lee, H., Hyun Yang, S., Bin Yu, C., Yeol Pak, S., Sung Oh, C., Seok Song, Y., June Kang, Y. & Ryoun Youn, J. 2017. Cell openness manipulation of low density polyurethane foam for efficient sound absorption. *Journal of Sound and Vibration*, 406, 224–236.

Ji, Y., Chen, S. & Cheng, Y. 2019. Synthesis and Acoustic Study of a New Tung Oil-Based Polyurethane Composite Foam with the Addition of Miscanthus Lutarioriparius. *Polymers*, 11, 1144.

Jin, F.-L., Zhao, M., Park, M. & Park, S.-J. 2019. Recent trends of foaming in polymer processing: A review. *Polymers*, 11, 953.

Kim, M.-S., Yan, J., Kang, K.-M., Joo, K.-H., Pandey, J. K., Kang, Y.-J. & Ahn, S.-H. 2013. Soundproofing properties of polypropylene/clay/carbon nanotube nanocomposites. *Journal of Applied Polymer Science*, 130, 504–509.

Kim, S., Le, T.-H., Choi, Y., Lee, H., Heo, E., Lee, U., Kim, S., Chae, S., Kim, Y. A. & Yoon, H. 2020. Electrical monitoring of photoisomerization of block copolymers intercalated into graphene sheets. *Nature Communications*, 11, 1324.

Kim, Y., Lee, S. & Yoon, H. 2021. Fire-safe polymer composites: Flame-retardant effect of nanofillers. *Polymers*, 13, 540.

Kim, Y. H. & Li, W. 2013. Multifunctional polyetherimide nanocomposite foam. *Journal of Cellular Plastics*, 49, 131–145.

Kuang, T., Chang, L., Chen, F., Sheng, Y., Fu, D. & Peng, X. 2016. Facile preparation of lightweight high-strength biodegradable polymer/multi-walled carbon nanotubes nanocomposite foams for electromagnetic interference shielding. *Carbon*, 105, 305–313.

Lee, J.-C., Hong, Y.-S., Nan, R.-G., Jang, M.-K., Lee, C. S., Ahn, S.-H. & Kang, Y.-J. 2009. Soundproofing effect of nano particle reinforced polymer composites. *Journal of Mechanical Science and Technology*, 22, 1468.

Lee, J., Kim, G. H. & Ha, C. S. 2012. Sound absorption properties of polyurethane/nano-silica nanocomposite foams. *Journal of applied polymer science*, 123, 2384–2390.

Lee, L. J., Zeng, C., Cao, X., Han, X., Shen, J. & Xu, G. 2005. Polymer nanocomposite foams. *Composites Science and Technology*, 65, 2344–2363.

Lercher, P. 2019. Combined transportation noise exposure in residential areas☆. *In:* Nriagu, J. (ed.) *Encyclopedia of Environmental Health (Second Edition)*. Elsevier.

Madaleno, L., Pyrz, R., Crosky, A., Jensen, L. R., Rauhe, J. C. M., Dolomanova, V., De Barros Timmons, A. M. M. V., Cruz Pinto, J. J. & Norman, J. 2013. Processing and characterization of polyurethane nanocomposite foam reinforced with montmorillonite–carbon nanotube hybrids. *Composites Part A: Applied Science and Manufacturing,* 44, 1–7.

Okolieocha, C., Raps, D., Subramaniam, K. & Altstädt, V. 2015. Microcellular to nanocellular polymer foams: Progress (2004–2015) and future directions – A review. *European Polymer Journal,* 73, 500–519.

Olszta, M. J., Cheng, X., Jee, S. S., Kumar, R., Kim, Y.-Y., Kaufman, M. J., Douglas, E. P. & Gower, L. B. 2007. Bone structure and formation: A new perspective. *Materials Science and Engineering: R: Reports,* 58, 77–116.

Park, J., Yang, S., Lee, H., Yu, C., Pak, S., Oh, C., Kang, Y. J. & Youn, J. 2017. Optimization of low frequency sound absorption by cell size control and multiscale poroacoustics modeling. *Journal of Sound and Vibration,* 397, 17–30.

Ray, S. S. 2013. *Clay-Containing Polymer Nanocomposites: from Fundamentals to Real Applications.* Newnes.

Rende, D., Schadler, L. S. & Ozisik, R. 2013. Controlling foam morphology of poly(methyl methacrylate) via surface chemistry and concentration of silica nanoparticles and supercritical carbon dioxide process parameters. *Journal of Chemistry,* 2013, 864926.

Si, J. Y., Tawiah, B., Sun, W. L., Lin, B., Wang, C., Yuen, A. C. Y., Yu, B., LI, A., Yang, W., Lu, H. D., Chan, Q. N. & Yeoh, G. H. 2019. Functionalization of MXene nanosheets for polystyrene towards high thermal stability and flame retardant properties. *Polymers (Basel),* 11, 18.

Sinha Ray, S. 2013. 11- Foam processing. *In:* SINHA RAY, S. (ed.) *Clay-Containing Polymer Nanocomposites.* Elsevier.

Song, P., Liu, B., Qiu, H., Shi, X., Cao, D. & Gu, J. 2021. MXenes for polymer matrix electromagnetic interference shielding composites: A review. *Composites Communications,* 24, 100653.

Sørensen, B. & Spazzafumo, G. 2018. 3 – Fuel cells. *In:* SØRENSEN, B. & SPAZZAFUMO, G. (eds.) *Hydrogen and Fuel Cells (Third Edition).* Academic Press.

Standau, T. & Altstädt, V. 2019. Foams. *In:* KARGER-KOCSIS, J. & BÁRÁNY, T. (eds.) *Polypropylene Handbook: Morphology, Blends and Composites.* Springer International Publishing.

Sundarram, S., Kim, Y. H. & Li, W. 2015. 4 –Preparation and characterization of poly(ether imide) nanocomposites and nanocomposite foams. *In:* MITTAL, V. (ed.) *Manufacturing of Nanocomposites with Engineering Plastics.* Woodhead Publishing.

Sung, C. H., Lee, K. S., Lee, K. S., Oh, S. M., Kim, J. H., Kim, M. S. & Jeong, H. M. 2007. Sound damping of a polyurethane foam nanocomposite. *Macromolecular Research,* 15, 443–448.

Svečko, R., Kusić, D., Kek, T., Sarjaš, A., Hančič, A. & Grum, J. 2013. Acoustic emission detection of macro-cracks on engraving tool steel inserts during the injection molding cycle using PZT sensors. *Sensors,* 13, 6365–6379.

Tan, Y.-J., Li, J., Tang, X.-H., Yue, T.-N. & Wang, M. 2020. Effect of phase morphology and distribution of multi-walled carbon nanotubes on microwave shielding of poly(l-lactide)/poly(ε-caprolactone) composites. *Composites Part A: Applied Science and Manufacturing,* 137, 106008.

Tripathy, D. P. 2008. *Noise Pollution.* APH Publishing.

Tyagi, M. & Tyagi, D. 2014. Polymer nanocomposites and their applications in electronics industry. *International Journal of Electronic and Electrical Engineering,* 7, 603–608.

Urbanczyk, L., Bourbigot, S., Calberg, C., Detrembleur, C., Jérôme, C., Boschini, F. & Alexandre, M. 2010. Preparation of fire-resistant poly(styrene-co-acrylonitrile) foams using supercritical CO2 technology. *Journal of Materials Chemistry,* 20, 1567–1576.

Wang, G., Li, W., Bai, S. & Wang, Q. 2019. Synergistic effects of flame retardants on the flammability and foamability of PS foams prepared by supercritical carbon dioxide foaming. *ACS Omega,* 4, 9306–9315.

Wang, L. K., Pereira, N. C. & Hung, Y.-T. 2005. *Advanced Air and Noise Pollution Control.* Springer.

Wu, G., Xie, P., Yang, H., Dang, K., Xu, Y., Sain, M., Turng, L.-S. & Yang, W. 2021. A review of thermoplastic polymer foams for functional applications. *Journal of Materials Science,* 56, 11579–11604.

Xie, Y., Li, Z., Tang, J., Li, P., Chen, W., Liu, P., Li, L. & Zheng, Z. 2021. Microwave-assisted foaming and sintering to prepare lightweight high-strength polystyrene/carbon nanotube composite foams with an ultralow percolation threshold. *Journal of Materials Chemistry C,* 9, 9702–9711.

Yao, D., Peng, N. & Zheng, Y. 2018. Enhanced mechanical and thermal performances of epoxy resin by oriented solvent-free graphene/carbon nanotube/Fe3O4 composite nanofluid. *Composites Science and Technology,* 167, 234–242.

Yuceturk, M. K., Abbasi, H., Antunes, M. D. S. P., Aydin, S. & Velasco Perero, J. I. Preparation of nanocomposite foams based on polysulfone and carbon-based nanoparticles using WVIPS. 19th International Metallurgy and Materials Congress, 25–27 October 2018, Istanbul, Turkey: congress proceedings book, 2018. 555–558.

Zhang, H.-B., Yan, Q., Zheng, W.-G., He, Z. & Yu, Z.-Z. 2011. Tough graphene– polymer microcellular foams for electromagnetic interference shielding. *ACS Applied Materials & Interfaces,* 3, 918–924.

Zhu, W., Chen, S., Wang, Y., Zhu, T. & Jiang, Y. 2018. Sound absorption behavior of polyurethane foam composites with different ethylene propylene diene monomer particles. *Archives of Acoustics,* 43, 8.

Zhu, Y., Luo, G., Zhang, R., Liu, Q., Sun, Y., Zhang, J., Shen, Q. & Zhang, L. 2019. Investigation of the constitutive model of W/PMMA composite microcellular foams. *Polymers,* 11, 1136.

8 An Overview about the Growing Usage of Elastomeric Foams and Nanocomposite Derivatives in the Foam Market

Ranimol Stephen

CONTENTS

8.1 INTRODUCTION

Polymeric foams have become an inalienable part of everyone's life. From the bed we sleep to the sofa which we relax on is all made of polymer foams. There might not be anyone who has not been touched by the utility of polymeric foams. As years go by we are seeing increased applications of polymeric forms in our daily lives though we might not be conscious about it. Polymeric foam is here to stay and will have a growing market. Recently, polymer nanocomposite-based foams are of great interest because of their low weight, low thermal conductivity, and good mechanical properties. Polymer nanocomposite foams can be used for spacecraft applications, firefighting equipment, insulation purposes, water treatment, and in the automobile industry as liners [1–3]. The growing demand for polymer foams arises due to their wide applications in various sectors. It is presented in Figure 8.1.

DOI: 10.1201/9781003218692-8

FIGURE 8.1   Applications of polymer foams.

8.2   MARKET OF POLYMER FOAMS AND THEIR NANOCOMPOSITES

Polymers for foam manufacturing are mainly polyethylene (PE), polypropylene (PP), phenolics, polyurethane (PU), cellulose acetate, urea-formaldehyde (UF), polyimides, silicones, etc., depending on the end-use applications. Also, foams are made of various elastomers such as chloroprene, butadiene, isoprene, acrylonitrile, Buna N, butyl rubbers, and others. The application of the final product dictates the choice of polymer for the manufacture of foams [4]. On comparing the properties of different polymeric foams phenolic foam has low thermal conductivity and good fire performance but it absorbs moisture. Elastomeric foam has low thermal conductivity, is flexible, and shows long-term performance while having poor mechanical properties. PE foams are inexpensive, low thermal conductivity; however, it is useful in a limited temperature range.

Extensive use of polymeric foams can cause environmental pollution; it is a major concern nowadays because of the dramatic climate changes. Most thermoplastic foams are recyclable and moldable while thermoset foams are not due to crosslinked structure. It is a versatile material that can be used for EMI shielding in this digital era. Due to the COVID-19 pandemic situation, the lifestyle of people is changed significantly, and they depend more on electronic gadgets for studies and work. Therefore, the need for lightweight EMI shields has increased over the years. Polymeric foams with conducting networks show good EMI shielding effectiveness [5,6]. Multiwalled carbon nanotubes incorporated in EPDM-based rubber composites exhibited EMI shielding efficiencies up to 45 dB (Figure 8.2). It is found to be a flexible and stretchable potential material as a wave absorber [7].

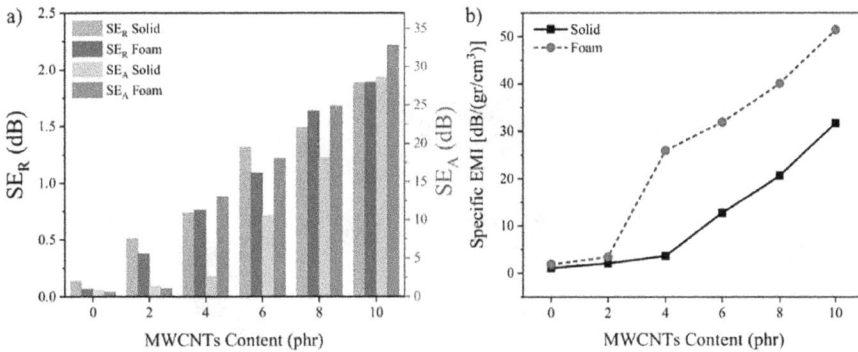

FIGURE 8.2 EMI shielding efficiency of EPDM/MWCNT (a) SE_A and SE_R at 10 GHz (Copy-right permission not required, Open Access article, MDPI) [7].

Flexible elastomeric foams market demand is mainly for thermal and acoustic insulation. According to the study of Fior market, the global market for elastomeric foams will be USD 2.91 billion by 2025 [8]. The driving factor in the growth of elastomeric foams is the heating, ventilation, and cooling (HVAC) systems in vehicles, buildings, etc. Foams provide an HVAC system and can prevent buildings or vehicles from heat during hot climatic conditions. Among elastomers, EPDM has more market because of its outstanding properties like low permeability, conformability, and insulation. It is widely used for thermal and acoustic insulation, and cushioning purposes in buildings and vehicles. Acrylonitrile butadiene rubber (NBR) is also used for making foams for HVAC systems. In brief, polymeric foams are used for energy-saving applications. As per the applications of foams in various sectors, the main market for foam is in the order HVAC system, automobile industry, electrical and electronics, and others [9]. The global market size of elastomeric foams is increased due to its extensive applications in various fields: cryogenic pipe insulation, refrigerant pipe insulation, domestic water pipe insulation, UV exposure and resistance, acoustic purposes, shock absorbers, and microwave absorbers.

Other than thermal and acoustic insulation, polymeric foams can be used for the treatment of water to remove heavy metal ions and oil//water separation. Polydimethyl siloxane (PDMS)- ZnSe elastomeric nanocomposite foam shows the removal of heavy metal ions from water through the cation exchange method. This method is found to be a faster technique for the efficient removal of heavy metal ions than other methods. Also, the color change in the foam indicates the presence of metal ions in the water. Therefore, it can be applicable for dual purposes, both removal and detection of heavy metal ions present in water (Figure 8.3) [10].

Different types of polymeric foams are used for the separation of oil/water and water/solvent from industrial wastes. PU foams are extensively used for thermal and acoustic applications owing to their ease of formation of microcellular structure and the tuning of pore size. Because of the same peculiarity, these foams are applicable in the separation of oil/water and oil/solvent [11–13]. The oil adsorption efficiency of PU foams depends on pores with the appropriate size and their high surface area.

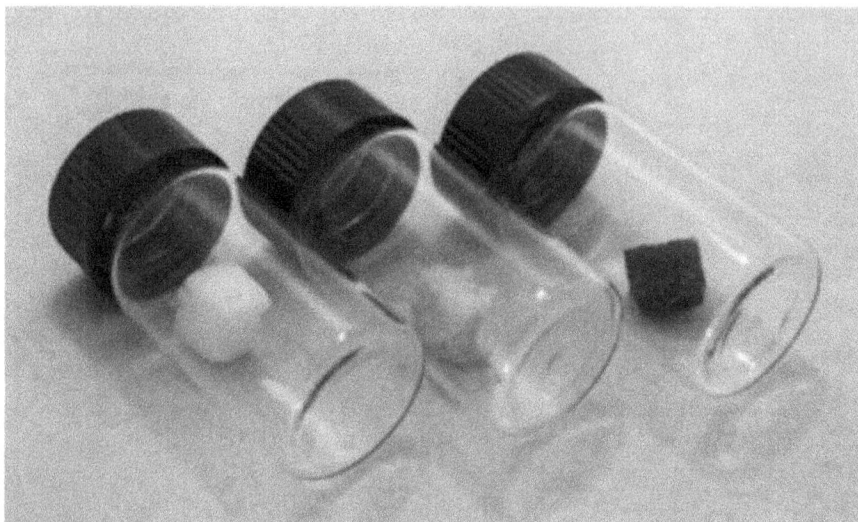

FIGURE 8.3 Photographs of PDMS elastomeric foam before (white), after ZnSe loading (yellow), and after the removal of Pb^{2+} ions from water (dark brown) (Reproduced with permission from ACS) [10].

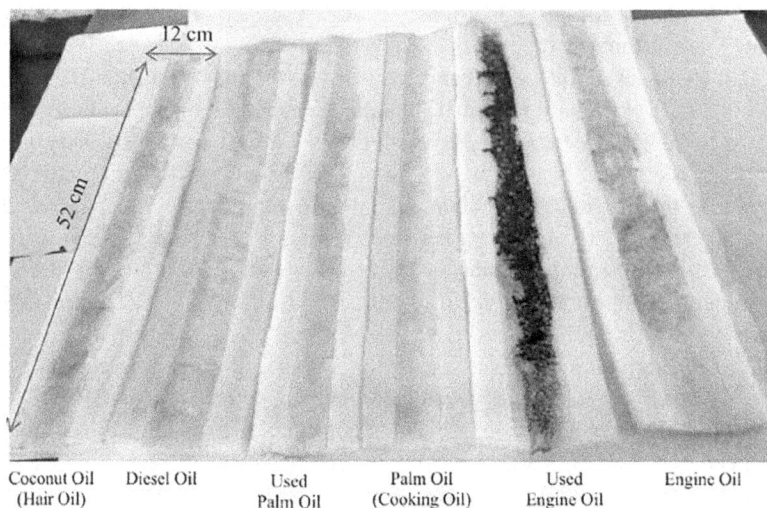

| Coconut Oil | Diesel Oil | Used | Palm Oil | Used | Engine Oil |
| (Hair Oil) | | Palm Oil | (Cooking Oil) | Engine Oil | |

FIGURE 8.4 Photographs of different types of oil absorbed on EPE polymeric foam after separation. (Copy-right permission not required, Open Access journal) [14].

Expanded PE (EPE) foams exhibited 78% oil absorption efficiency from the oil/water mixture. The overall study demonstrated the capability of EPE foam in absorbing oil from oil-water mixtures with a 78% absorption efficacy (Figure 8.4) [14].

Polydopamine functionalized copper foam shows oil absorption up to 98%, also possesses good mechanical strength and low shrinkage [15]. Polytetrafluoroethylene (PTFE) coated Ni foam also exhibited remarkably good oil recovery [16]. Other potential materials proved for oil-water separation are PDMS [17,18], PS [19,20], and PVDF [21,22] foams. Kandasubramanian and co-workers reviewed the application of various foams in oil/water separation applications, their toxic effect on the environment, and the factors affecting the effective oil/water separation [23].

8.3 REASON FOR THE GROWTH OF THE FOAM MARKET

Polymer foams are considered to be an important porous material. By incorporating certain additives such as nanofillers, one can improve their properties for extensive applications like piezoresistive material, EMI shielding, and microwave absorbers rather than thermal and acoustic insulation. Another advantage is the possibility of tuning the pore structure by controlling certain parameters as per their end-use applications.

The main reasons behind the growing market of polymeric foam are (i) lightweight, (ii) reasonably flexible, (iii) good heat and sound insulation, and (iv) corrosion-resistant. Consequently, rise in demand for foam in the construction, automobile, furniture, footwear, packaging, sports, and recreation segments. Conservation of energy drives the market of foams as they can be used in HVAC systems. Moreover, few polymers and their nanocomposite foams are used for water treatment. Depending on the surface area, pore size, presence of certain additives, and functional groups, foams exhibited high oil and heavy metal ion absorption efficacy.

8.4 CHALLENGES AND FUTURE OUTLOOK IN THE FOAM INDUSTRY

Disposability is one of the major challenges faced by the foam industry. Waste management, recycling, remolding, emission of toxic chemicals on burning and different chemicals used during processing also affect the growth rate of foam industry. Proper safety measures have to be adopted to mitigate these problems because they will be resulting in severe environmental pollution. Even though polymeric foams have extensive applications in various sectors, the industry has to tackle the problems to avoid environmental pollution. These issues retard the growth of the foam market despite its demand in various segments. Temporarily, the outburst of COVID-19 also influences the market of foam. Because it is commonly used in the construction, transportation, furniture, sports, recreational, and packaging areas. Two areas are still to be explored: (i) the wide applications of polymer foam nanocomposites, and (ii) the usage of waste foams in water treatment.

REFERENCES

1. W. Jiang, S.S. Sundarram, D. Wong, J.H. Koo, W. Li, Polyetherimide nanocomposite foams as an ablative for thermal protection applications, *Composites Part B*, 2014, 58, 599, doi:10.1016/j.compositesb.2013.10.040.
2. A. Vahidifar, E. Esmizadeh, E. Rostami, S.N. Khorasani, D. Rodrigue, Morphological, rheological, and mechanical properties of hybrid elastomeric foams based on natural rubber, nano clay, and nanocarbon black, *Polymer Composites*, 2019, 40(11), 4289, doi: 10.1002/pc.25290.

3. A. Kumar, B. Patham, S. Mohanty, S.K. Nayak, Polyolefinic nanocomposite foams: Review of microstructure-property relationships, applications, and processing considerations, *Journal of Cellular Plastics*, 2020, doi: 10.1177/0021955X20979752.

4. K.C. Khemani, *Polymeric Foams: An Overview, Polymeric Foams ACS Symposium Series*; American Chemical Society: Washington, DC, 1997.

5. Y. Zhan, M. Oliviero, J. Wang, A. Sorrentino, G.G. Buonocore, L. Sorrentino, M. Lavorgna, H. Xia, S. Iannace, Enhancing the EMI shielding of natural rubber-based supercritical CO_2 foams by exploiting their porous morphology and CNT segregated networks, *Nanoscale*, 2019, 11, 1011, doi: 10.1039/C8NR07351A.

6. J. Yang, X. Liao, J. Li, G. He, Y. Zhang, W. Tang, G. Wang, G. Li, Light-weight and flexible silicone rubber/MWCNTs/Fe_3O_4 nanocomposite foams for efficient electromagnetic interference shielding and microwave absorption, *Composites Science and Technology*, 2019, 181, 107670, doi: 10.1016/j.compscitech.2019.05.027.

7. H. Bizhani, A.A. Katbab, E. Lopez-Hernandez, J.M. Miranda, R. Verdejo, Highly deformable porous electromagnetic wave absorber based on ethylene–propylene–diene monomer/multiwall carbon nanotube nanocomposites, *Polymers*, 2020, 12, 858, doi: 10.3390/polym12040858.

8. Fior markets, Flexible Elastomeric Foam Market by Type (Natural Rubber/Latex, Chloroprene, Others), Function, End-User Industry, Regions, Global Industry Analysis, Market Size, Share, Growth, Trends, and Forecast 2018 to 2025, 2019. https://www.fiormarkets.com/report/flexible-elastomeric-foam-market-by-type-natural-rubber-latex-385923

9. Elastomeric Foam Market Size, Share & COVID-19 Impact Analysis, By Function (Thermal Insulation, Acoustic Insulation), By Type (EPDM, Nitrile Rubber (NBR), Others), By Application (HVAC, Automotive & Transportation, Electrical & Electronics, Others), and Regional Forecast, 2020-2027, October 2020. http://www.fortunebusinessinsights.com/elastomeric-foam-market-104004.

10. A.A. Chavan, H. Li, A. Scarpellini, S. Marras, L. Manna, A. Athanassiou, D. Fragouli, Elastomeric nanocomposite foams for the removal of heavy metal ions from water, *ACS Applied Materials Interfaces*, 2015, 7, 27, 14778, doi: 10.1021/acsami.5b03003.

11. V. Nandwana, S.M. Ribet, R.D. Reis, Y. Kuang, Y. More, V.P. Dravid, O.H.M. Sponge, A versatile, efficient, and ecofriendly environmental remediation platform, *Industrial & Engineering Chemistry Research.*, 2020, 59, 10945, doi: 10.1021/acs.iecr.0c01493.

12. J. Joy, J. Abraham, J. Sunny, J. Mathew, S.C. George, Hydrophobic, super absorbing materials from reduced graphene oxide/MoS2 polyurethane foam as a promising sorbent for oil and organic solvents, *Polymer Testing*, 2020, 87, 106429, doi: 10.1016/j.polymertesting.2020.106429.

13. L.S. Martins, N.C. Zanini, L.S. Maia, A.G. Souza, R.F.S. Barbosa, D.S. Rosa, D.R. Mulinari, Crude oil and S500 diesel removal from seawater by polyurethane composites reinforced with palm fiber residues, *Chemosphere*, 2021, 267, 129288, doi: 10.1016/j.chemosphere.2020.129288.

14. C.S. Patil, V.R. Patil, S.N. Anbhule, C.J. Khilare, G.B. Kolekar, A.H. Gore, Waste packaging polymeric foam for oil-water separation: An environmental remediation, *Data in Brief*, 2018, 19, 86, doi: 10.1016/j.dib.2018.05.033.

15. W. Zhou, G. Li, L. Wang, Z. Chen, Y. Lin, A facile method for the fabrication of a superhydrophobic polydopamine-coated copper foam for oil/water separation, *Applied Surface Science*, 2017, 413, 140, doi: 10.1016/j.apsusc.2017.04.004.

16. K.Y. Eum, I. Phiri, J.W. Kim, W.S. Choi, J.M. Ko, H. Jung, Superhydrophobic and superoleophilic nickel foam for oil/water separation, *Korean Journal of Chemical Engineering*, 2019, 36, 1313, doi: 10.1007/s11814-019-0308–9.

17. E. Piperopoulos, L. Calabrese, A. Khaskhoussi, E. Proverbio, C. Milone, Thermophysical characterization of carbon nanotube composite foam for oil recovery applications, *Nanomaterials*, 2020, 2020, 10, doi: 10.3390/nano10010086.

18. S.-J. Choi, T.-H. Kwon, H. Im, D.-I. Moon, D.J. Baek, M.-L. Seol, J.P. Duarte, Y.-K. Choi, A polydimethylsiloxane (PDMS) sponge for the selective absorption of oil from water, *ACS Applied Materials & Interfaces*, 2011, 3, 4552, doi: 10.1021/am201352w.

19. Y. Liu, K. Zhang, Y. Son, W. Zhang, L.M. Spindler, Z. Han, L. Ren, A smart switchable bioinspired copper foam responding to different pH droplets for reversible oil–water separation, *Journal of Materials Chemistry A* 2017, 5, 2603, doi: 10.1039/C6TA10772A.

20. L. Yu, G. Hao, L. Xiao, Q. Yin, M. Xia, W. Jiang, Robust magnetic polystyrene foam for high efficiency and removal oil from water surface, *Separation and Purification Technology*, 2017, 173, 121, doi: 10.1016/j.seppur.2016.09.022.

21. R. Bentini, A. Pola, L.G. Rizzi, A. Athanassiou, D. Fragouli, A highly porous solvent free PVDF/expanded graphite foam for oil/water separation, *Chemical Engineering Journal*, 2019, 372, 1174, doi: 10.1016/j.cej.2019.04.196.

22. R. Arora, K. Balasubramanian, Hierarchically porous PVDF/nano-SiC foam for distant oil-spill cleanups, *RSC Advances*, 2014, 4, 53761, doi: 10.1039/c4ra09245g.

23. K.V. Udayakumar, P.M. Gore, B. Kandasubramanian, Foamed materials for oil-water separation, *Chemical Engineering Journal Advances*, 2021, 5, 100076, doi: 10.1016/j.ceja.2020.100076.

9 Polymeric Foams and Their Nanocomposite Derivatives for Shock Absorption

Behrad Koohbor and George Youssef

CONTENTS

9.1 INTRODUCTION

Foams are an important class of materials for impact mitigation and shock absorption applications, as evident from their widespread utilization in several industrial sectors prone to impact and shock loadings [1,2]. Many solid polymers have been transformed into cellular structures (i.e., foamed); nonetheless, there are ongoing and sustained research efforts in foaming more polymeric materials with unique microstructures and with the addition of nano-reinforcement constituents. Polyurethane and polystyrene are the most common in industrial applications susceptible to impact and shock loading conditions, such as packaging sensitive parcels (e.g., electronic devices, scientific instrumentations) during shipping and sports protective gears such as shin guards and shoulder pads [3]. They are also used in the automotive industry to mitigate the bluntness of frontal or rear impacts during traffic accidents. For example, automobile bumpers are lined with rigid polyurethane foam to improve the impact protection of cars due to fender bender collisions. In sports gear, expanded polystyrene foams are found in many devices, including American Football helmets, to manage and absorb the impact energy while improving comfort and fit. Moreover, a significant weight reduction is a byproduct of using polymeric foams

DOI: 10.1201/9781003218692-9

for improving the impact performance, leading to a substantial increase in other performance metrics such as fuel economy in vehicles and reducing inertial forces in biomechanical collisions [2]. Hence, the integration of polymeric foams entails primary benefits (e.g., impact mitigation or shock absorption), secondary advantages (e.g., weight reduction), and tertiary improvements (e.g., improved power-to-weight ratio) [4–6]. Polymeric foams are also ubiquitous in construction insulations [7]. The multifaceted benefits mark the uniqueness and potential opportunities associated with polymeric foams.

At its essence, foams are porous or cellular materials, where the solid polymeric materials (rigid or flexible) form the overall structure, entrapping relatively large volume fractions of gaseous or liquid substances. Figure 9.1 shows a scanning electron microscope (SEM) micrograph of a typical cellular microstructure of polymeric foams, demonstrating the significantly large unoccupied regions while the solid portion is entrapped in the cell walls. Therefore, and in the context of the current topic, the primary efficacy of polymeric foams in managing the impact energy hinges on the three essential and symbiotic factors that naturally define or describe this class of materials, encompassing the properties of the base polymer (material), the attributes of the cellular structure (geometry), and the interaction between entrapped and enclosing substances (fluid-solid interactions). Nanocomposites have recently emerged as a novel method to improve the performance of the foams using nanoscale-sized and shaped particulates to tune the stiffness, strength, and toughness of the resulting foams [8–10]. Arguably, these parameters also strongly depend on the manufacturing method (e.g., extrusion vs. mold casting) and the conditions surrounding the fabrication process. This process-structure-property-performance nexus is at the core of materials science and engineering and subsequently at play here as well. More broadly, accounting for the object to be protected, the energy-absorbing efficiency of foams also depends on:

1. The form factor (i.e., the shape and geometry of different features defining this object);
2. The characteristics of the foam itself (e.g., the ability to dissipate impact energy); and
3. The temporal characteristics delineating the tolerance of the object to impact [1].

The impact tolerance of the foam has implications on the thickness, geometry, and density of the foam. In all, the efficacy of the foam to absorb the impact energy depends on the impact direction with respect to the preferred axis of optimal properties (if any), the dissipation mechanisms at play during the impact event, including elastic, plastic, or brittle deformations of the foam microstructure, the flow of entrapped fluid, and the presence of reinforcement mechanisms or constituents [1,2].

Before embarking on more details, while metallic foams have been studied and integrated into some impact and shock-loading applications [11,12], the keen focus here is on polymers as the base material. The energy dissipated by a foam under mechanical load is generally quantified as the area under the plateau region of the stress-strain curve, which captures several energy absorption mechanisms, such as

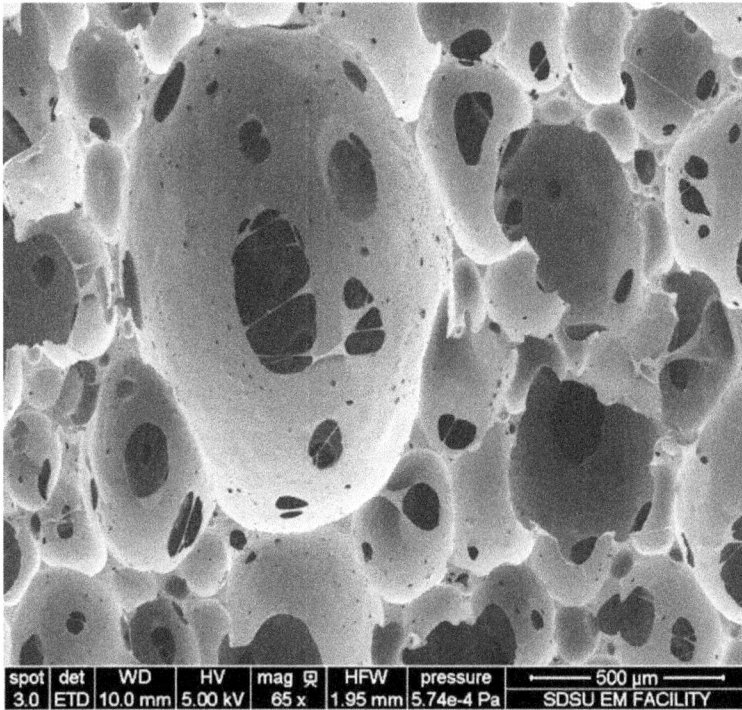

spot	det	WD	HV	mag ⚲	HFW	pressure	⟵ 500 µm ⟶
3.0	ETD	10.0 mm	5.00 kV	65 x	1.95 mm	5.74e-4 Pa	SDSU EM FACILITY

FIGURE 9.1   SEM micrograph of a non-rigid foam, showing a perforated-spherical cellular microstructure (picture from the author's current research collaboration).

the cellular scale instabilities, including cell buckling (elastic and plastic), yielding, and crushing. The plateau region dominates the stress-strain response of polymeric foams due to limited linear elasticity at the onset of loading and excessive material nonlinearities in the densification region at the outset. As a result, the plateau region accounts for significant energy loss at a nearly constant load. Figure 9.2 is an example stress-strain curve of elastomeric foam showing the effect of density on the mechanical performance [13]. With this in mind, there are three general mechanistic attributes of polymers making them a very attractive class of materials for impact mitigation and shock resistance applications, namely, hyperelasticity (excessive elasticity), viscoelasticity (time-dependent elasticity), and manufacturability (e.g., net-shape or near-net-shape processing) [14]. Elastomeric and ductile polymers, exhibiting glass transition below room temperature, experience large reversible deformations, i.e., hyperelasticity, giving rise to substantially higher energy dissipation than other material systems. The hyperelastic response is clearly desirable since it extends the plateau region and increases the dissipated impact energy. Furthermore, the time-dependent properties of the base polymers unleash another source of energy dissipation due to the inherent dampening attributes that arise from the lag between the input force and induced deformation. The contribution of the irrecoverable energy from the viscoelastic hysteresis to the overall efficacy of the form in absorbing the impact energy

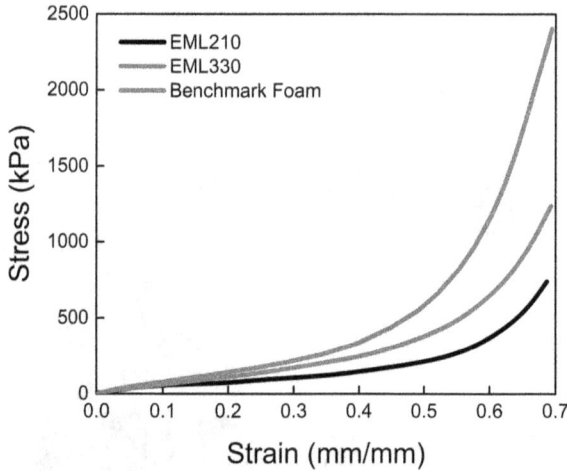

FIGURE 9.2 Example stress-strain curves of elastomeric foams with hybrid open/closed cellular structure (EML210 has a nominal density of 210 kg/m^3 while EML330 has a nominal density of 330 kg/m^3) compared to a closed-cell foam with a nominal density of 397 kg/m^3 [13].

hinges on the rate sensitivity of the base polymer and its congruence to the loading rate. Finally, polymeric foams can be molded into complex shapes, essentially inheriting the intricate details of the object to be protected. A prime example of this is the two-part polyurethane foam used to protect large electronic and scientific instrumentations during shipping, poured in the shipping container or box to ensure proper fitment and protection (i.e., poured-in-place foaming process). Mechanistically, the energy dissipation of foam is an explicit function of the stress (and deformation) while being implicit of strain rate and temperature during the impact event [14].

To increase the functionality and improve the properties of foamed polymers, a relatively small amount of microscale or nanoscale particles can be added to the based polymer during fabrication, resulting in microcomposites and nanocomposites, respectively [15,16]. In the form of platelets or fibers, the reinforcement phase exhibits drastically different physical properties (e.g., mechanical, electrical, thermal) than the base polymers, improving the performance at a minimal weight penalty. In nanocomposite foams, at least one of the dimensions has to be at the nanoscale while the others can be at the same scale (e.g., clay, silica, graphite, metals, copolymers nanoparticles) or several microns (such as nanotubes and nanofibers, e.g., single-walled or multi-walled carbon nanotubes) [16–22]. The filler phase was reported to substantially improve the impact absorption, mechanical stiffness and strength, thermal insulation, and electrical and thermal conductivity [23]. An additional byproduct of the reinforcement particulates includes the alteration of cellular structure and foaming nucleation. The reinforcement filler can be used to alter the structure by adjusting the size of the cell, ranging from the macroscale (conventional foams) to microscale and nanoscale foams with, for example, nanocellular geometries in the range of 0.1–100 nm [15,16], irrespective of the cell morphology being either open or

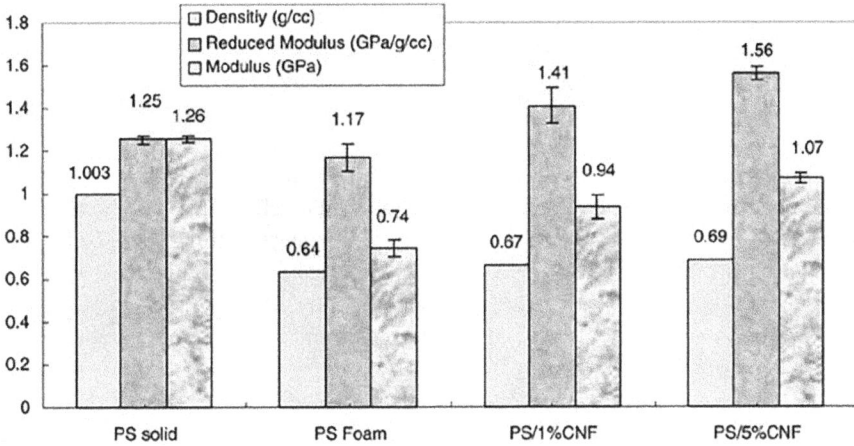

FIGURE 9.3 Nanocomposites of polystyrene foam with carbon nanofibers, improving the modulus at low weight fraction [15]. Reproduced with permission from Elsevier, 2005.

closed cell. The latter is desirable for thermal insulation application, while the former might be more suitable for impact mitigation situations. The viability of manufacturing nanocomposite foams hinges on the distribution and dispersion of the nanoparticles into the base polymer, where the physical attributes and chemical properties of the nanoscale fillers play a crucial role in the fabrication process. Generally, nanocomposites foams can be accomplished via solution blending (solvent-suspended nanoparticles dispersed in the polymer melt), melting blending (dispersion of solvent-free nanoparticles into the polymer melt), or *in situ* polymerization (nanofillers are mixed with the monomer before polymerization commences) processes [24–31]. Several research studies reported nanocomposite foams based on different polymers, including polyurethane, polystyrene, polyvinyl chloride, polyethylene oxide, polyvinyl alcohol, and polyacrylic acid, to name a few [15]. Figure 9.3 shows the effect of nanofiller on the modulus and density of polystyrene foams, compared to the properties of the bulk polymer and conventional polystyrene counterparts, where carbon nanofibers (CNFs) at 1% and 5% content were investigated. The reduced modulus (i.e., specific stiffness = modulus/density) increased from 1.17 GPa/g/cm^3 for conventional polystyrene foam to 1.41 GPa/g/cm^3 and 1.56 GPa/g/cm^3 for polystyrene foam with 1% and 5% CNFs, respectively [32]. The modulus increased nearly 45% by the mere addition of 5% CFNs. This example demonstrates the technological superiority of foamed nanocomposites, an area of great research potential.

9.2 STRUCTURAL HIERARCHY IN FOAMS

9.2.1 Definition of Length Scales

The mechanical behavior of polymeric foams is directly correlated with their structural hierarchy. The term structural hierarchy refers to the interrelations between the base (parent) polymer and the geometry and architecture of the cell structure and the macroscopic properties that arise from such interrelations. Therefore, to understand the macroscale properties of foams, we must realize the correlations between material behaviors at multiple length scales, from micro to meso to macro.

In most general terms, polymeric foams consist of two main phases: an interconnected solid polymer network and a gaseous phase. The empty channels occupied by the gas can be either interconnected (in open-cell foams), completely separated (in closed-cell foams), or partly connected (in semi-closed cell foams, like those shown in Figure 9.1). Aside from the fact the degree of openness within the cell structure attributes to properties at larger scales, it is imperative to first define the aforementioned length scales in terms of size, connectivity, and mechanical behaviors. Figure 9.4 illustrates a general description of length scales and the material characteristics/ properties associated with them.

The microscale behavior in polymeric foams is referred to as physical length scales typically covering 10^{-6}–10^{-4} m dimensions [33]. The mechanical behavior of the base polymer is the governing factor at such small length scales. In other words, the microscale behavior of the foam is essentially controlled by the properties of the

Macroscale

Global stress-strain behavior
Stiffness
Poisson's ratio
Strain rate sensitivity

Mesoscale

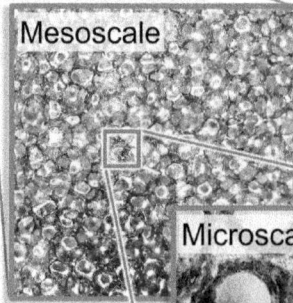

Cell architecture
Representative volume element
Solid-gas interactions
...

Microscale

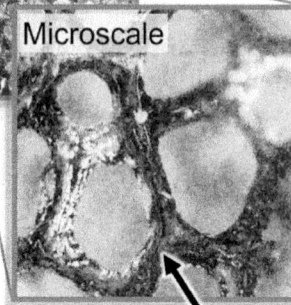

Chemistry
Microstructure
Reinforcement particles
...

Base material

FIGURE 9.4 Illustration of various length scales and their associated properties in atypical open-cell foam.

material that constitutes the solid parts in a foam cellular network. As such, certain phenomena and mechanisms that contribute to the mechanical behavior of the parent polymer are of great importance at microscales. Examples of such phenomena are the chemistry and the constituent phases inside the polymer as well as the degree of material heterogeneity. Furthermore, the elastic, hyperelastic, and viscoelastic of the base polymer contribute to the overall behavior, as discussed in the previous section. The addition of nanofillers to enhance the properties of the foams at larger scales has the most substantial effect at such micro and sub-microscales [15].

On the opposite side of the length-scale spectrum is the macroscopic scale (macroscale). At the macroscale, the behavior of foam is characterized by its overall bulk properties. The notion of macroscale behavior implies that the foam sample must be large enough to encompass a sufficient number of individual cells so that the bulk properties can be estimated by homogenizing (or smearing) the material behavior at smaller length scales. In other words, there is a minimum number of cells that must be considered such that the homogenized and collective behaviors of the cells will be representative of the bulk properties of the foam. This minimum number of cells, also referred to as the statistical representative volume element (SRVE), has been characterized for several different foams using both experimental and numerical methods. In a recent experimental effort, the minimum number of cells needed to satisfy the SRVE requirements was shown to be approximately 4,000, identified for a hyperelastic polyurea foam with ~100 kg/m^3 nominal density [34]. Compared with the average cell size of the examined foam, this number was found to be equivalent to a ~3 mm physical length scale. Nonetheless, for practical purposes and most engineering foams, macroscale can be referred to as dimensions equal to or above 10^{-2} m dimensions.

Finally, the mesoscopic scale (mesoscale) in foams and other cellular structures is defined as a transitional length scale that bridges the micromechanics behavior of foams to their bulk macroscale properties. Unlike the other two length scales described above, there are no strict rules to define the physical dimensions of mesoscales in foams. Instead, mesoscale dimensions are most often referred to as those ranging from the size of an individual cell to a collection of cells that set the lower limits of the macroscale behavior. This length scale is typically in the range of 10^{-4}–10^{-2} m. The most important point to note when considering the material behavior at mesoscales is the fact that the foam behavior at this scale is strongly dependent on the cells shape, geometry, and connectivity, as well as the interactions between the solid material entrapped in the cell walls and the fluid (often a gas) that occupies the cell interiors. Therefore, identifying the upper limits of the mesoscale in foams is of immense importance. The upper limits of the mesoscales (which also set the lower limits of the macroscales in foams) are generally regarded as the physical dimensions of the material's SRVE. The SRVE concept in foams and other cellular structures is essential as it is required for a proper determination of sample sizes required for accurate testing of the foam. In the absence of standard protocols, testing samples whose dimensions are smaller than the SRVE size of the foam can lead to under/over-estimating of the bulk properties. In the following, we focus on some of the more important properties of foams and their correlations with the structural hierarchy in the material.

9.2.2 PROPERTY-STRUCTURE RELATIONSHIPS

Perhaps the single most important parameter that affects the macroscopic behavior of foams is the relative density, ρ^*. This parameter is simply defined as the ratio between the foam density (ρ_f) and the base materials density (ρ_s), expressed as:

$$\rho^* = \frac{\rho_f}{\rho_s} \tag{9.1}$$

The relative density of a foam is correlated with the shape, morphology, strut thickness, cell size, and the connectivity of cells in a foam. Mathematical formulae have been developed to estimate the relative density of foams. The interested reader is referred to Ref. [1] to learn more about cellular structure-relative density formulations for a variety of foams and other cellular solids.

Almost all bulk mechanical properties of foams are strongly dependent on relative density. For example, numerous mathematical formulae have been developed to correlate the mechanical properties of foams and other cellular structures to their relative density. Although developed based on simplified mechanics of materials concepts, these expressions have proven extremely useful in addressing some of the fundamental concepts regarding the physical, mechanical, failure, and rate-sensitive performances of foams. For example, it has been demonstrated that rigid foams are susceptible to internal damage and loss of stiffness when subjected to shock and impact loads significantly smaller than the macroscale yield/failure stress of the foam. The mechanisms associated with such internal damage formations were revealed in rigid polyurethane foams by correlating global stresses with relative density (see Figure 9.5). The relative density in this figure is indicated by the ratio between the cell wall thickness, t, and cell diameter, l.

Another interesting example of property-structure relationships was reported in [13], wherein the density-dependent stiffness, plateau stress, and densification strain in a novel semi-closed cell flexible polyurea foam (illustrated in Figure 9.1) were estimated using density-dependent mathematical formulae expressed for open and closed-cell foams. Structural features of the foam were used along with the mechanical properties of the base polymer to predict the stress-strain responses of the foam at various relative densities. A mathematical expression correlating the relative density, properties of the based polymer, and the entrapped gas pressure effect with the effective modulus (E_s) of the novel polyurea foams was developed as:

$$\frac{E_{eq}^*}{E_s} = \frac{C_1^2 \Phi^2 \rho^{*3} + C_1 C_2 (1 - \Phi) \rho^{*2} + C_1 \dfrac{p_0 \left(1 - 2v_f\right) \rho^*}{E_s \left(1 - \rho^*\right)}}{C_1 \rho^* (1 - \Phi^2) + C_2 (1 - \Phi) + \dfrac{p_0 \left(1 - 2v_f\right)}{E_s \left(1 - \rho^*\right) \rho^*}} \tag{9.2}$$

where Φ denotes the fraction of solid material contained in the cell walls, p_0 is the gas pressure inside the cells, and v_f is the apparent Poisson's ratio of the foam,

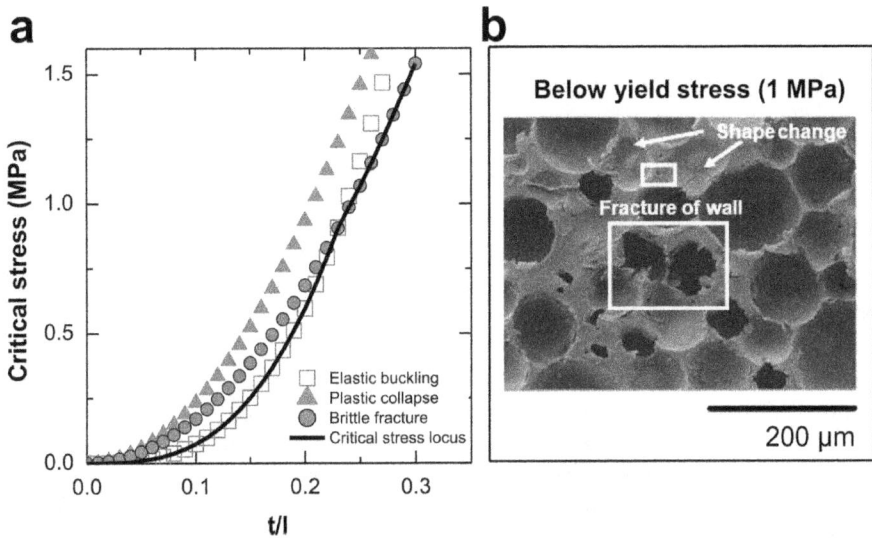

FIGURE 9.5 (a) Variation of critical stress (global compressive stress associated with failure at cellular scales) with respect to relative density for the three most probable failure mechanisms: elastic buckling, plastic collapse, and brittle failure. (b) Evidence of the presence of microscopic damage occurred at impact stresses significantly lower than the apparent yield stress of the material. Data reproduced from Ravindran *et al.* [35] with permission from Elsevier, 2016.

determined experimentally. Parameters C_1 and C_2 are proportionality constants that can be measured by simple compression tests.

Similar structure-property correlations were developed for novel hyperelastic polyurea foams to describe the role of internal pressure build-up and the solid-gas interactions when the foam is subjected to impact loading [36]. An example of such correlations is discussed in detail in Section 9.4.3, highlighting the importance of relative density (i.e., the contribution of structural characteristics), base material's Poisson's ratio (i.e., the contribution of the material response), internal gas pressures (i.e., the role of processing method), and strain-rate dependent strength, as the resultant property.

9.3 MECHANICS PRELIMINARIES ON IMPACT AND SHOCK LOADING

The diverse utilization of foams in impact mitigation applications stipulates a broad range of loading rates based on the velocity of the impacting projectile (potential energy to kinetic energy conversion), where the impact velocity can be quantified in a relatively straightforward way, e.g., $v_i = \sqrt{2gh}$ for a free-fall situation. This impact velocity is the initial velocity at the onset of the impact event, which is augmented by the velocity due to the gravitational acceleration and the velocity due to the deceleration of the projectile as it deforms the foam protective pad or cushion under investigation. In such a free-fall situation, the velocity-time history, $v(t)$, can be calculated as:

$$v(t) = v_i + gt - \int_0^t \frac{F(t)}{m} dt \qquad (9.3)$$

where $F(t)$ is the impact force-time signal and m is the mass of the impacting projectile. The third term in Eq. 9.3 is the contribution of the acceleration (or deceleration) based on Newton's Second Law. The impact-induced deformation due to the impact velocity and the foam structure dimensions define the strain rate. The impact-induced deformation can be calculated by integrating Eq. 9.3 and treating the impacting projectile as a rigid member that experiences negligible deformation with respect to the foam deformation. This calculation is discussed below based on the force balance. To demonstrate this concept, let us consider dropping a projectile from 1m height, resulting in ~4.4m/s initial impact velocity (i.e., the first term in Eq. 9.3), which is translated to a strain rate in the order of 44 s⁻¹ based on a 0.1 m thickness of a foam pad (compare to 0.05 /s strain rate in quasi-static testing, e.g., stress-strain curves shown in Figure 9.2). In general, the nominal strain rate ($\dot{\varepsilon}$) due to the impact is

$$\dot{\varepsilon} = \frac{v}{l_o} \qquad (9.4)$$

where $v(t)$ is the impact velocity and l_o is the initial thickness of the impacted foam pad. Thus, increasing the impact velocity is associated with elevated strain rates that may supersede the speed of sound in the foam, giving rise to additional inertial effects that result in rapidly growing plateau stresses [37]. In laboratory testing situations, the impact velocity can be increased via two approaches: (1) increasing the drop height, or (2) using a velocity-controlled testing machine, where the projectile is accelerated before the drop [36]. The first approach relies on manipulating the potential energy, while the latter hinges on increasing the kinetic energy.

Finally, shock (or ballistic) loading is associated with strain rates ranging between 10^3 and 10^5 s⁻¹, where the dynamic behavior of the base material, the geometry of the cellular structure, and the entrapped fluid play crucial roles in the impact efficacy of the foam, as discussed in the forthcoming sections. The high strain loading can be accomplished using the Split Hopkinson bar experiment or flay-plate impact test [33,35,38,39]. The next section is dedicated to the interrelationship between the strain rate and the foam attributes, including the material, microcellular structure, and the working fluid within the foam cells. The latter has a pronounced effect at ultra-high strains at levels comparable to the densification strains.

The preceding discussion was based on laboratory testing, posing an important question: *Is there a need to investigate the foam response to high strain rate?* The answer is unequivocal, Yes! In the previous paragraph, we discussed a simple example of a drop weight from a height of one meter, resulting in a strain rate of ~40 s^{-1}. However, other impact scenarios can result in much higher strain rates. For example, a passenger car driving at a speed of 40 km/h experiences a strain rate of 200 s^{-1} [40] due to the collision, while the energy-absorbing foams in the side panels can achieve strain rates up to 1,500 s^{-1} if the vehicle is driving at 90 km/h [40]. Hence, studying the high strain rate and shock responses of polymeric foams, regardless of the size of the cellular structure (macro or nano) and the materials (thermoset or thermoplastic), is imperative for developing proper protective foams. While low-strain (quasi-static) testing has been used to forecast the dynamic response of foams [2,13,36,41,42], these approaches neglect significant dynamic effects. The quasi-static-based analyses do not account for the strong strain rate effect on the base materials (i.e., polymers in this case) and the interrelationship between the deformation modes and characteristics of the dynamic loading. The latter submit the foam to localized evidential deformations and higher stresses than low strain rate conditions, resulting in drastically different failure modes. The deformation localization is exaggerated if the cellular structure is at the nanoscale because of geometrical nonlinearities [43,44], or due to the difference in the mechanical properties (e.g., modulus) in the presence of a reinforcement phase [44]. Recently, a new approach was introduced to enhance the performance by a self-reinforcing mechanism, where microspheres of the same materials (not to be confused with secondary reinforcing materials, as discussed above) nucleated during the foaming process. Since the reinforcement phase is made of the same material, the interfacial bonding and property disparity issues were eliminated [45]. More research is needed to explicate the effect of self-reinforcement on the dynamic response of this type of elastomeric foams, especially at ultrahigh strain rate loading scenarios.

The answer to the question stated above, while worthy, is very challenging due to the ambiguous interplay between the geometry and distributions of the cells, and the dynamic properties of the material, including the strain rate sensitivity and attributes of the impact scenario. Such ambiguities are hyperbolic in the case of nanocomposite foams due to the distribution of the reinforcement within the micro or nano cellular structure of the foam. For example, Luong et al. [46] studied the strain rate effect on polyvinyl chloride foams and found that the yield strength increases as a function of strain rate, i.e., resembling the behavior of the bulk polymer but at a different rate. Figure 9.6 exemplifies the effect of strain rate, ranging from 10^{-4} to 2×10^3 s^{-1}, on the mechanical performance of polymeric foams [46], demonstrating the evolution in the modulus, plateau region, and onset of densification. Ramachandra showed that the energy absorption of closed-cell foam increased substantially when the impact velocity exceeded a certain threshold (10 m/s in their case) while being insensitive to the velocity below the threshold [47]. Consistent with the findings of Ramachandra, Chen and collaborators found that at strain rates exceeding 113 s^{-1}, the compressive strength of expanded polystyrene foam highly depended on the strain rate [48]. Since the microcellular structure and the base material play a crucial role, Sun et al. [33] postulated that different localized deformation and stresses are present throughout the sample during loading. Collectively, understanding the effect of strain rate on the overall behavior is vital for the development of a superior impact mitigation structure; however, more research is needed to explicate the mechanisms leading to the

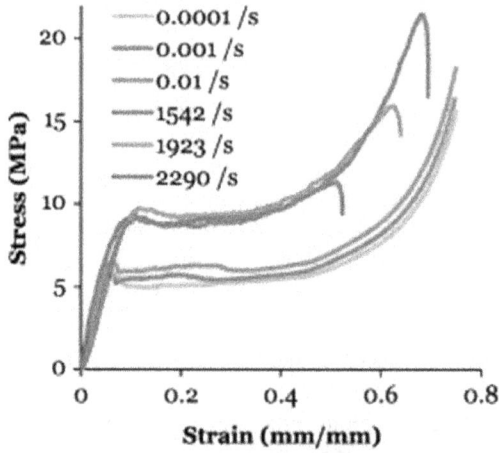

FIGURE 9.6 Compressive stress-strain curves of closed-cell Divinycell PVC foam with a nominal density of 200 kg/m³ at quasi-static and high strain rate [46]. Reproduced with permission from Elsevier, 2013.

source a threshold above which the strain rate effect becomes pronounced. Some of the more important mechanisms governing the strain rate effects in the high-rate loading of foams are discussed in Section 9.4.

The objective of mechanical testing, irrespective of the strain rate, is to deduce the stress-strain behavior, which can be used to develop impact-tolerant structures based on the requirements stipulated by the application. Mills compiled a list of case studies [2], demonstrating a practical approach to the development of foam-based protective structures that nicely supplements the prior seminal work by Gibson and Ashby [1]. The reader is referred to these titles for comprehensive coverage of phenomenological and micromechanical models describing the behavior of polymeric foams and their utility in practical applications. The focus here is reviewing the basics of high-strain loadings, including the mechanics of drop impact and Split Hopkinson bar, leading to the extractions of the stresses and strains from the measurements.

Let us consider a drop impact experiment (shown schematically in Figure 9.7), where a mass is dropped from a specific height on a deformable foam sample. The force balance based on Newton's law implies that the exerted force is the sum of the forces due to deceleration and gravity, resulting in an expression for the acceleration, a, as:

$$a = \frac{F(t)}{m} - g \qquad (9.5)$$

where $F(t) = A \times \sigma(t)$ with A denoting the surface area of the impacting projectile and $\sigma(t)$ being the stress imparted on the sample due to the impact. Here, $F(t)$ is the foam restraining force, which is measured using a load cell or force sensor installed between the impacting projectile and the sample. Rearranging Eq. 9.5 to solve for the velocity (recall that $a = \dfrac{dv}{dt}$) and integrating the rearranged expression leads to the velocity expression in Eq. 9.3 above. Moreover, we can use the definition of the strain rate to write an expression for the strain rate based on the preceding discussion:

$$\dot{\varepsilon} = \frac{v_i}{l_o} + \frac{1}{ml_o} \int_0^t F(t)\,dt - \frac{gt}{l_o}. \qquad (9.6)$$

Sequentially, integrating Eq. 9.6 and solving for the strain results in:

$$\varepsilon = \frac{v_i}{l_o}t + \frac{1}{ml_o} \int_0^t \int_0^t F(t)\,dt\,dt - \frac{gt^2}{2l_o}. \qquad (9.7)$$

Therefore, using the measured force and Eq. 9.7 above yields the dynamic stress-strain curve, which can then be used to evaluate the foam performance based on mechanics metrics, including ideality, G-level, and efficiency of the foam, as demonstrated in [13,49].

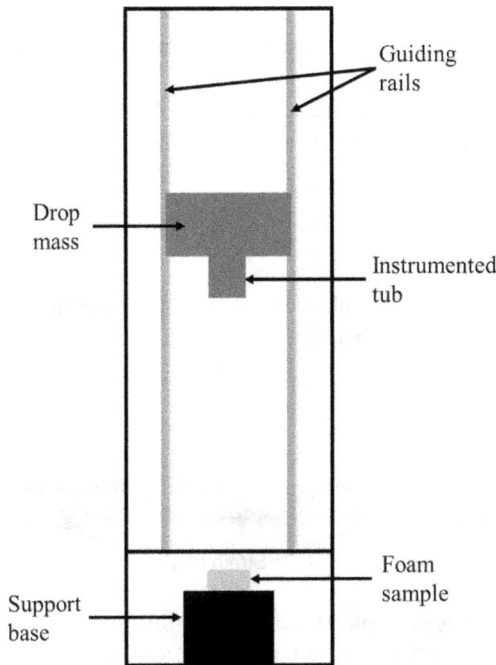

FIGURE 9.7 Schematics of drop weight impact testing setup and a typical force-time history recording due to the impact event.

In the case of the split Hopkinson pressure bar (SHPB) experiment for high strain rate testing, the strain is measured using strain gauges integrated as part of the setup. Figure 9.8 shows a simplified representation of an SHPB setup, where a striker bar is actuated using a gas gun (sudden release of pressurized gas), inducing a stress wave in the incident bar. The incident bar submits the sample to the high strain rate loading, later transferring the stress wave to the transmission bar. Two strain gauges are attached to the incident and transmission bars, respectively, and the signal is collected using high-speed oscilloscope or data acquisition system. In the SHPB experiment, the average strain rate and strains applied on the test can be determined using the following equations [50]:

$$\dot{\varepsilon}(t) = -\frac{2C_b}{l_o} \varepsilon_r(t) \qquad (9.8)$$

$$\varepsilon(t) = \int_0^t \frac{2C_b}{l_o} \varepsilon_r(t) dt \qquad (9.9)$$

where ε_r denotes the reflected strain, i.e., the signal collected by the strain gage on the incident bar and upon the reflection of the wave from the first bar-sample interface. C_b is the velocity of the sound wave in the bar, which can be readily determined as the square root of the ratio between the bar material's elastic modulus and mass density (i.e., $C_b = \sqrt{\frac{E}{\rho}}$). The stress history generated in the foam sample can then be determined using the following simple equation that correlates the temporal variation of stress to the transmitted strain signal, ε_t, as:

$$\sigma(t) = \frac{E_b A_b}{A_s} \varepsilon_t(t) \qquad (9.10)$$

where E_b, A_b, and A_s denote the elastic modulus of the bar material, cross-sectional area of the bar, and the load-bearing cross-sectional area of the foam test piece, respectively.

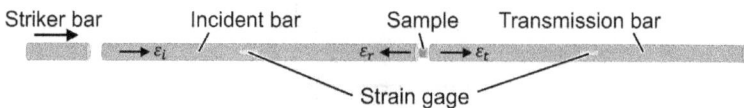

FIGURE 9.8 Schematics of a split Hopkinson bar (SHPB) used for high strain rate testing of materials. ε_i, ε_r, and ε_t denote incident, reflected, and transmitted strain signals, respectively.

9.4   STRAIN-RATE EFFECTS: PROPERTY-MICROSTRUCTURE-PERFORMANCE RELATIONSHIPS

Owing to their excellent impact mitigation capacities, foams are widely used in applications where a combination of good load-bearing and energy absorption at reduced structural weights is required. For example, foams have been ideal for the core material in sandwich panels and other hollow structures, used primarily as protective components against blast and impact loads. The enhanced energy absorption capacity of foams under impact loading originates from their low mechanical impedance. Such low mechanical impedance characteristics can be traced down to the microstructural features, such as the cellular architecture and entrapped gas effects in foams.

The low mechanical impedance behavior in foams is associated with the low-stress wave speed in these materials. When subjected to impact or shock loading conditions, the low elastic wave speed in foams causes a delayed establishment of stress equilibrium in the structure [51]. In addition, the slow formation and propagation of compaction waves (developed in rigid foams under impact loading conditions) lead to the generation of highly nonuniform deformation fields in a foam structure, as shown in Figure 9.9. While such delayed equilibrium conditions make impact testing and characterization of foams challenging, they also make foams an ideal class of materials for mechanical energy absorption and damping purposes.

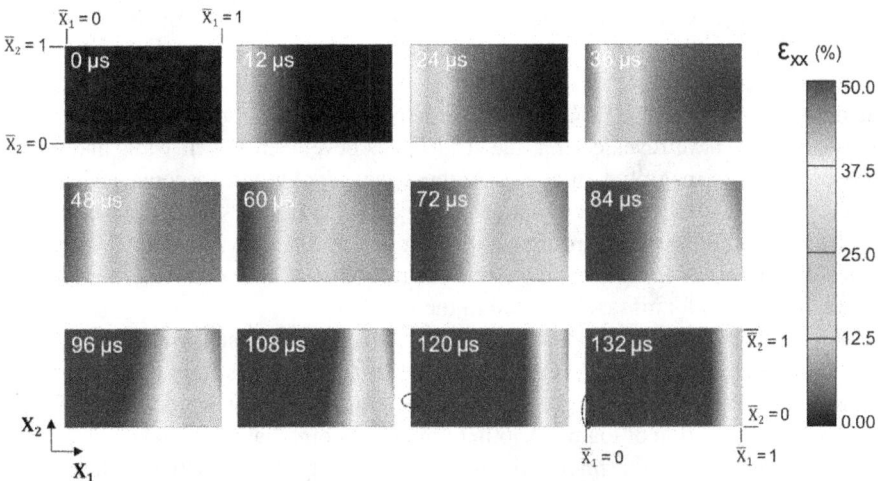

FIGURE 9.9   Development of axial strain fields (along X_1 direction) in a rigid polyurethane foam subjected to impact loading. The impact load has been applied on the left end (i.e., $\bar{X}_1 = 0$). The temporal evolution of strain fields also indicates the propagation of the compaction band front along the sample [35]. The sample in this work is a rigid polyurethane foam with a nominal density of 154 kg/m³. Reproduced with permission from Elsevier, 2018.

Irrespective of the fundamental mechanisms that attribute to the non-equilibrium stress states in foams, these materials show exceptionally high strain rate-dependent mechanical properties. The strain rate-dependent behavior in foams is manifested in their moduli, Poisson's ratios, yield and plateaus stresses, densification strains, and all energy absorption metrics.

Strain rate dependence of foams, in general, has been correlated with three inter-dependents phenomena that can be activated at various strain rate conditions. Originated from the microstructural features in foams, the three main sources of high strain rate sensitivity in foams are: strain rate-dependent behavior of the base material, rate-dependent release of entrapped gas, and the effect of micro-inertia at micro and mesoscales. These phenomena are briefly discussed in the following.

9.4.1 EFFECT OF BASE MATERIAL

Polymer foams are essentially created as a network of polymeric struts that are inter-connected in a random (stochastic) or semi-ordered fashion. Regardless of the cell structure and geometry, the base polymer in foams can be a rigid or flexible polymer with or without a reinforcement phase. Polymers show a higher degree of strain rate sensitivity compared with other classes of materials, e.g., most metals and ceramics (for more information, the interested reader is referred to Siviour *et al.* [52], wherein a comprehensive overview of the strain rate effects in polymers is discussed). When polymer foams are subjected to elevated strain rates, the applied deformations are carried by the solid polymer regions in the structure. Microscopic strain rates exerted on solid struts may be vastly different from those applied on the foam piece at global scales. There exist modeling studies based on simplified cell geometries that suggest microscopic strain rates one order of magnitude smaller than those applied at macroscopic scales [53]. On the other hand, recent *in situ* experimental strain measurements have revealed that the strain rates developed in cell walls and struts can be orders of magnitude higher than those measured at macroscopic scales [54]. Interestingly, such significant variations between the strain rates developed at the two scales are found to be most prominent at lower strain values, as shown in Figure 9.10. The latter observation signifies the effect of micro-inertia as another source of strain rate sensitivity in foams, as discussed in the following.

Although there are still debates as to how the strain rate sensitivity of the base material correlates with that of the foam, it is reasonable to assume that the contribution of the base polymer must be taken into consideration when dealing with the high-rate deformation of foams. Another important point that needs to be considered with extreme care is the microstructural and cellular characteristics of the foam, as they significantly contribute to the strain rate sensitivity of foams, in general. Nevertheless, the assumption of the contribution of the base material in the strain rate sensitivity of the foams has been experimentally verified, at least for foams with a random cell structure.

9.4.2 INERTIA EFFECTS IN MICRO AND MACROSCALES

The low mechanical impedance behavior in polymeric foams gives rise to significant inertia effects during high strain rate loading [38,39,55]. The development of

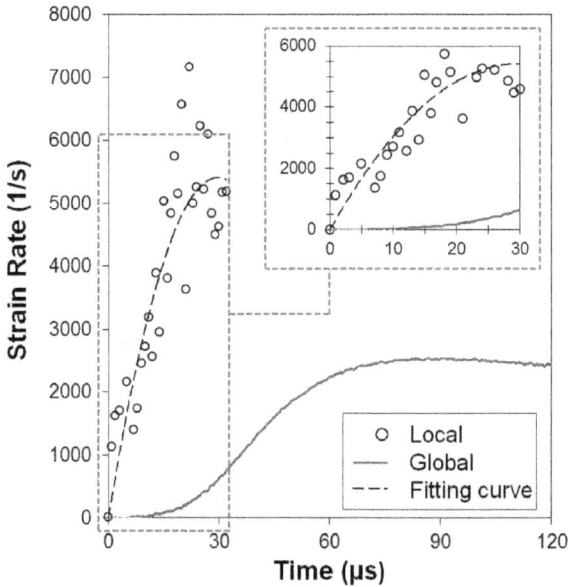

FIGURE 9.10 A Comparison between local (on cell struts) and global (developed over the entire test piece) strain rates in a foam piece subjected to uniaxial impact. The sample in this work is a rigid polyurethane foam with a nominal density of 560 kg/m³. Strain rates developed at the length scales were measured using high-speed digital image correlation [54]. Reproduced with permission from Elsevier, 2018.

high hydrostatic stresses manifests the inertia effects in the high strain rate loading of foams at macroscales during the early stages of deformation. The development of these macroscale inertia effects is a direct consequence of the low elastic wave speeds in foams, simply due to their low modulus-to-density ratios. However, the high strain rate loading of foams also involves the development of *micro-inertia* effects, which are different from the inertia effects at macroscopic length scales.

Micro-inertia effects in impacted foams are generated due to the acceleration of the solid struts and cell walls in the material [33,56]. Similar to the effect of strain rate at different length scales, the acceleration fields developed on solid struts in a foam substantially affect the mechanical properties of the foam, especially at early deformation times. Also similar to the strain rate effect, micro-inertia effects are sensitive to cellular and microstructural characteristics in foams. It has been shown that the role of micro-inertia in foams fabricated from polymers with higher densities is more prominent than their low-density counterparts. In addition, structures in which cell walls are not perfectly aligned with the direction of the applied global load are more susceptible to higher micro-inertia effects. To explain the latter statement, one should note that any misalignment between the direction of an applied global load and the cell walls inside the foam can lead to the resistance of cell walls against lateral deformation. The resistance to deformation at microscales is amplified at global scales by a substantial virtual hardening behavior demonstrated by the stress-strain behavior of the material.

Analyses of the inertia effects at multiple length scales have been conducted in a number of research articles in the past few years. Some of the analytical works on the topic have been supplemented by *in-situ* characterizations facilitated by state-of-the-art measurement techniques. The interested reader is referred to references [33] for more in-depth insights into the role of inertia in the strain-rate sensitive behavior of foams and other cellular solids.

9.4.3 EFFECT OF ENTRAPPED GASSES

Another factor that attributes to the strain rate sensitivity of foams in dynamic loading conditions is the rate-dependent release of the entrapped gas from within the cell structure of the material. The effect of entrapped gas has been more thoroughly studied for closed and semi-closed cell foams since the intra-cell pressurization effect is more prominent when the gas release is hindered by the closed (or semi-closed) cell walls. Previous studies supported by both experimental and computational modeling results suggest that the entrapped gas hardly affects the load-bearing and stress development behaviors in the material in small deformations [57]. Modeling predictions have also pointed to the relatively unaffected plateau stresses due to the internal gas release effect. On the other hand, the entrapped gas effect has been shown to have a more substantial effect on the densification strain in closed-cell foams. The latter is extremely important, as the densification onset strain in foams is considered an essential measure for its impact energy mitigation behavior.

The strength enhancement in foams due to the internal gas effect has been quantified by simplified analytical models. The simplified assumptions used for the development of predictive mathematical tools needed to study the entrapped gas effect are primarily associated with the structure and connectivity of cells. For example, the entrapped gas effects in 2D cellular structures with repeated hexagonal and circular cells have been studied in [57]. The simplifying assumptions of regular cell shapes (either circular or hexagonal) facilitate a straightforward approach for the quantification of the relative density of the foam, as it is a crucial piece of information in determining the pressure buildup in cells during mechanical loading. The entrapped gas pressurization process can then be modeled using the assumption of an ideal gas that can be pressurized under isothermal or adiabatic conditions. Accordingly, the stress increase due to the pressurized entrapped gas (denoted by $\Delta\sigma$) can be determined at dynamic loading conditions by the equation,

$$\Delta\sigma = \left\{ \frac{1-\rho^*}{1-\rho^*-(1-2\upsilon)\varepsilon} \left[\left(\frac{\dot{\varepsilon}_{a-c}-\dot{\varepsilon}}{\dot{\varepsilon}_{a-c}-\dot{\varepsilon}_{i-c}} \right) + \left(\frac{\dot{\varepsilon}-\dot{\varepsilon}_{i-c}}{\dot{\varepsilon}_{a-c}-\dot{\varepsilon}_{i-c}} \right) \left(\frac{1-\rho^*}{1-\rho^*-(1-2\upsilon)\varepsilon} \right)^{\gamma-1} \right] -1 \right\} p_0$$

(9.11)

where $\dot{\varepsilon}$ denotes the applied strain rate, $\dot{\varepsilon}_{i-c}$ and $\dot{\varepsilon}_{a-c}$ are the critical strain rates associated with the isothermal and adiabatic conditions with values equal to the 1 s^{-1} and 100 s^{-1}, respectively. υ and γ indicate the apparent Poisson's ratio of the foam and the heat capacity ratio of air, respectively. p_0 is the initial pressure of the gas inside the material. A series of analytical modeling efforts based on the above equations and complemented by finite element analyses suggest that the stress increase due to the internal pressure buildup only occurs at large strain conditions, most closely associated with the densification onset strains. In addition, the entrapped gas effect becomes more prominent at higher strain-rate conditions.

In recent studies, similar analytical approaches were used to investigate the effect of gas pressurization in flexible foams [36]. In an effort to understand the internal gas effect, the relative density of the foam was replaced by experimentally measured strain-dependent values. It was then shown that compared with other factors, the entrapped gas effect plays a less significant role, especially when the applied strain rates are on the order of a few 100 s^{-1}. Instead, it was revealed that the geometry, connectivity, and the degree of openness of the cells in a semi-closed flexible foam have more substantial effects on the strain rate sensitivity of the material.

Characterizing the effects of entrapped gas in closed-cell foams has been a topic of research for several decades. In spite of numerous computational and numerical models developed to study this effect, experiments and investigations on the topic have been scarce. Nevertheless, there exist studies that investigate the problem using creative approaches. An example of such creative experimental studies is the mechanical tests performed in a fluid chamber [40]. Careful quantification of the gas bubbles that escaped from the surface of the foam has been used as a metric to quantify the gas release rate as a function of strain rate and strain applied on the foam piece.

Despite its challenges, understanding the strain-rate sensitivity in foams has attracted tremendous attention in the past few decades. Following the recent trends in the production of foams with superior load-bearing and energy absorption capacities, research studies on the topic have also ramped up. Recent developments in the fabrication of novel foams, especially those with auxetic cell structures and graded densities [58], demand for more in-depth investigations to uncover the property-microstructure-performance relationships. It is, therefore, anticipated that the coming years will witness a growing interest in the development of novel characterization approaches that will be helpful in understanding the underlying mechanisms that govern the strain rate-dependent behavior of next-generation foams and other cellular solids at multiple times and length scales.

9.5 SUMMARY

This chapter provided a brief overview of polymer foams (and their nanocomposite derivatives) as an ideal class of materials that offer excellent shock and impact absorption characteristics along with a tunable load-bearing response, both achievable at significantly reduced structural weights. These unique properties are only achievable through a careful design of the material and structural features at multiple length scales, from micro and sub-micro to macroscopic dimensions. The significance of structural hierarchy was highlighted through the definition of length scales and their effect on the apparent properties of foams. Detailed discussions were provided on the role of strain rate, how it affects the behavior of foams at various scales, and practical methodologies developed for the application of controlled high strain rate loading conditions on foam test pieces. Attempts were made to highlight the most important and general aspects of the structure-property correlations in foams, especially in response to high rate and shock loading conditions, although there still exist numerous case-specific phenomena that may govern the response of foams in various loading scenarios. Nevertheless, the authors believe that the present document provides the reader with a comprehensive overview of both fundamental and practical information on the shock loading behavior of foams in the most general sense. Through the incorporation of a rich list of references on the topic, the authors have attempted to refer interested readers to other useful resources for a more in-depth study of the relevant topics.

REFERENCES

1. L. J. Gibson, M. F. Ashby. *Cellular Solids – Structure and Properties (2nd ed.)*, Cambridge University Press, 1999.
2. N. Mills. *Polymer foams handbook: Engineering and biomechanics applications and design guide*, Elsevier, 2007.
3. T. Khan, V. Acar, M. Raci Aydin, B. Hulgu, H. Akbulut, M. O. Seydibeyglu. A review on recent advances in sandwich structures based on polyurethane foam cores. *Polymer Composites* 2020; 41:2355–2400.
4. K. O'leary, K. A. Vorpahl, B. Heiderscheit. Effect of cushioned insoles on impact forces during running. *Journal of the American Podiatric Medical Association* 2008; 98 (1):36–41.
5. L. Blanc, A. Jung, S. Diebels, A. Kleine, M. O. Sturtzer. Blast wave mitigation with galvanised polyurethane foam in a sandwich cladding. *Shock Waves* 2021; 31:525–540.
6. M. Asad, T. Zahra, D. P. Thambiratnam, T. H. T. Chen, Y. Zhunge. Geometrically modified auxetic polyurethane foams and their potential application in impact mitigation of masonry structures. *Construction and Building Materials* 2021; 311:125170.
7. L. Aditya, T. M. I. Mahlia, B. Rismanchi, H. M. Ng, M. H. Hasan, H. S. C. Metselaar, O. Muraza, H. B. Aditiya. A review on insulation materials for energy conservation in buildings. *Renewable and Sustainable Energy Reviews* 2017; 73:1352–1365.
8. F. Jin, M. Zhao, M. Park, S-J. Park. Recent trends of foaming in polymer processing: A review. *Polymers* 2019; 11(6):953.
9. C. Okolieocha, D. Raps, K. Subramaniam, V. Alstadt. Microcellular to nanocellular polymer foams: Progress (2004–2015) and future directions–A review. *European Polymer Journal* 2015; 73:500–519.
10. L. Chen, D. Renden L. S. Schadler, R. Ozisik. Polymer nanocomposite foams. *Journal of Materials Chemistry A* 2013; 1(12):3837–3850.
11. T. Wan, Y. Liu, C. Zhou, X. Chen, Y. Li. Fabrication, properties, and applications of open-cell aluminum foams: A review. *Journal of Materials Science & Technology* 2021; 62:11–24.
12. B. H. Smith, S. Szyniszewski, J. F. Hajjar, B. W. Schafer, S. R. Arwade. Steel foam for structures: A review of applications, manufacturing and material properties. *Journal of Constructional Steel Research* 2012; 71:1–10.
13. N. Reed, N. U. Huynh, B. Rosenow, K. Manlulu, G. Youssef. Synthesis and characterization of elastomeric polyurea foam. *Applied Polymer Science* 2020; 137:48839.
14. G. Youssef, *Applied Mechanics of Polymers Properties, Processing, and Behavior*, 1st Edition, Elsevier, 2021, ISBN: 9780128210789.
15. L. J. Lee, C. Zeng, X. Cao, X. Han, J. Shen, G. Xu. Polymer nanocomposite foams. *Composites Science and Technology* 2005; 65:2344–2363.

16. V. Mittal, *Polymer Nanocomposite Foams*, 1st Edition, CRC Press, 2014, ISBN: 9781138074996.
17. R. Verdejo, R. Stampfli, M. Alvarez-Lainez, S. Mourad, M. A. Rodriguez-Perez, P. A. Bruhwiler, M. Shaffer. Enhanced acoustic damping in flexible polyurethane foams filled with carbon nanotubes. *Composites Science and Technology* 2009; 69(10):1564–1569.
18. E. Rezvanpanah, S. R. GhaffarianAnbaran, E. Di maio. Carbon nanotubes in microwave foaming of thermoplastics. *Carbon* 2017; 125:32–38.
19. R. Verdejo, C. Saiz-Arroyo, J. Carretero-Gonzalez, F. Barroso-Bujans, M. A. Rodriguez-Perez, M. A. Lopez-Manchado. Physical properties of silicone foams filled with carbon nanotubes and functionalized graphene sheets. *European Polymer Journal* 2008; 44(9):2790–2797.
20. S. Liu, R. Eijkelenkamp, J. Duvigneau, G. J. Vancso. Silica-assisted nucleation of polymer foam cells with nanoscopic dimensions: impact of particle size, line tension, and surface functionality. *ACS Applied Materials & Interfaces* 2017; 9(43):37929–37940.
21. S. Liu, S. Yin, J. Duvigneau, G. J. Vancso. Bubble seeding nanocavities: multiple polymer foam cell nucleation by polydimethylsiloxane-grafted designer silica nanoparticles. *ACS Nano* 2020; 14(2):1623–1634.
22. I. Javni, W. Zhang, V. Karajkov, Z. S. Petrovic, V. Divjakovic. Effect of nano-and micro-silica fillers on polyurethane foam properties. *Journal of Cellular Plastics* 2020; 38(3):229–239.
23. S. Liu, J. Duvigneau, G. J. Vancso. Nanocellular polymer foams as promising high performance thermal insulation materials. *European Polymer Journal* 2015; 65:33–45.
24. L. Liu, Z. Qi, X. Zhu. Studies on nylon 6/clay nanocomposites by melt-intercalation process. *Journal of Applied Polymer Science* 1999; 71:1133–1138.
25. D. M. Lincoln, R. A. Vaia, Z.-G. Wang, B. S. Hsiao. Secondary structure and elevated temperature crystallite morphology of nylon-6/layered silicate nanocomposites. *Polymer* 2000; 42(4):1621.
26. H. R. Dennis, D. L. Hunter, D. Chang, S. Kim, J. L. White, J. W. Cho, D. R. Paul. Effect of melt processing conditions on the extent of exfoliation in organoclay-based nanocomposites. *Polymer* 2001; 42(23):9513–9522.
27. J. W. Cho, D. R. Paul. Nylon 6 nanocomposites by melt compounding. *Polymer* 2001; 42:1083–1094.
28. T. D. Fornes, P. J. Yoon, H. Keskkula, D. R. Paul. Nylon 6 nanocomposites: the effect of matrix molecular weight. *Polymer* 2001; 42(25):09929–09940.
29. P. H. Nam, P. Maiti, M. Okamoto, T. Kotaka, N. Hasegawa, A. Usuko. A hierarchical structure and properties of intercalated polypropylene/clay nanocomposites. *Polymer* 2001; 42(23):9633–9640.
30. G. Galgali, C. Ramesh, A. Lele. A rheological study on the kinetics of hybrid formation in polypropylene nanocomposites. *Macromolecules* 2001; 34(4):852–858.
31. V. V. Ginzburg, A. C. Balazs. Predicting the phase behavior of polymer-clay nanocomposites: the role of end-functionalized chains, ACS Symposium Series. *Polymer Nanocomposites* 2002; 804:57–70.
32. J. Shen, X. Han, L. J. Lee. *Nucleation and reinforcement of carbon nanofibers on polystyrene nanocomposite foam. 63rd Edition. Annual Technical Conference. Society of Plastics Engineers*; 2005.
33. Y. Sun, Q. M. Li. Dynamic compressive behaviour of cellular materials: A review of phenomenon, mechanism and modelling. *International Journal of Impact Engineering* 2018; 112:74–115.
34. B. Koohbor, N. Pagliocca, G. Youssef. A multiscale experimental approach to characterize micro-to-macro transition length scale in polymer foams. *Mechanics of Materials* 2021; 161:104006.

35. S. Ravindran, B. Koohbor, P. Malchow, A. Kidane. Experimental characterization of compaction wave propagation in cellular polymers. *International Journal of Solids and Structures* 2018; 139–140:270–282.
36. B. Koohbor, A. Blourchian, K. Z. Uddin, G, Youssef. Characterization of energy absorption and strain rate sensitivity of a novel elastomeric polyurea foam. *Advanced Engineering Materials* 2021; 23:2000797.
37. B. Koohbor, N. K. Singh. Radial and axial inertia stresses in high strain rate deformation of polymer foams. *International Journal of Mechanical Sciences* 2020; 181:105679.
38. B. Koohbor, A. Kidane, W.-Y. Lu, M. A. Sutton. Investigation of the dynamic stress–strain response of compressible polymeric foam using a non-parametric analysis. *International Journal of Impact Engineering* 2016; 91:170–182.
39. B. Koohbor, A. Kidane, W.-Y. Lu. Effect of specimen size, compressibility and inertia on the response of rigid polymer foams subjected to high velocity direct impact loading. *International Journal of Impact Engineering* 2016; 98:62–74.
40. R. Bouix, P. Viot, J.-L. Lataillade. Polypropylene foam behaviour under dynamic loadings: Strain rate, density and microstructure effects. *International Journal of Impact Engineering* 2009; 36:329–342.
41. K. C. Rusch. Load-compression behavior of flexible foams. *Journal of Applied Polymer Science* 1969; 13:2297–2311.
42. J. Miltz, G. Gruenbaum. Evaluation of cushioning properties of plastic foams from compressive measurements. *Polymer Engineering and Science* 1981; 21(15):1010–1014.
43. S. Costeux. CO_2 blown nanocellular foams. *Journal of Applied Polymer Science* 2014; 131(23): 1–16.
44. G. Luo, Y. Zhu, R. Zhang, P. Cao, Q. Liu, J. Zhang, Y. Sun, H. Yuan, W. Guo, Q. Shen, L. Zhang. A Review on Mechanical Models for Cellular Media: Investigation on Material Characterization and Numerical Simulation. *Polymers* 2021; 13(19):3283.
45. S. Do, N. U. Huynh, N. Reed, A. M. Shaik, S. Nacy, G. Youssef. Partially-perforated self-reinforced polyurea foams. *Applied Sciences* 2020; 10(17):5869.
46. D. D. Luong, D. Pinisetty, N. Gupta. Compressive properties of closed-cell polyvinyl chloride foams at low and high strain rates: Experimental investigation and critical review of state of the art. *Composites Part B: Engineering* 2013; 44:402–416.
47. S. Ramachandra, P. S. Kumar, U. Ramamurty. Impact energy absorption in an Al foam at low velocities. *Scripta Materialia* 2003; 49:741–745.
48. W. Chen, H. Hao, D. Hughes, Y. Shi, J. Cui, Z.-X. Li. Static and dynamic mechanical properties of expanded polystyrene. *Materials & Design* 2015; 69:170–180.
49. M. Avalle, G. Belingardi, R. Montanini. Characterization of polymeric structural foams under compressive impact loading by means of energy-absorption diagram. *International Journal of Impact Engineering* 2001; 25:455–472.
50. S. Ravindran, A. Tessema, A. Kidane. Local deformation and failure mechanisms of polymer bonded energetic materials subjected to high strain rate loading. *Journal of Dynamic Behavior of Materials* 2016; 2:146–156.
51. B. Koohbor, S. Mallon, A. Kidane, W-Y. Lu. The deformation and failure response of closed-cell PMDI foams subjected to dynamic impact loading. *Polymer Testing* 2015; 44:112–124.
52. C. R. Siviour, J. L. Jordan. High strain rate mechanics of polymers: A review. *Journal of Dynamic Behavior of Materials* 2016; 2:15–32.
53. V. S. Deshpande, N. A. Fleck. High strain rate compressive behaviour of aluminium alloy foams. *International Journal of Impact Engineering* 2000; 24:277–298.
54. B. Koohbor, S. Ravindran, A. Kidane. Effects of cell-wall instability and local failure on the response of closed-cell polymeric foams subjected to dynamic loading. *Mechanics of Materials* 2018; 116:67–76.

55. B. Song, B. Sanborn, W.-Y. Lu. Radial inertia effect on dynamic compressive response of polymeric foam materials. *Experimental Mechanics* 2019; 59:17–27.
56. Z. Lovinger, C. Czarnota, S. Ravindran, A. Molinari, G. Ravichandran. The role of micro-inertia on the shock structure in porous metals. *Journal of the Mechanics and Physics of Solids* 2021; 154:104508.
57. Y. Sun, Q. M. Li. Effect of entrapped gas on the dynamic compressive behaviour of cellular solids. *International Journal of Solids and Structures* 2015; 63:50–67.
58. O. Rahman, K. Z. Uddin, J. Muthulingam, G. Youssef, C. Shen, B. Koohbor. Density-graded cellular solids: Mechanics, fabrication, and applications. *Advanced Engineering Materials* 2022; 24:2100646. https://doi.org/10.1002/adem.202100646

10 An Overview on Nanoparticles, Nanocomposites and Emerging Applications of Polymeric Nanocomposite Foams

Aswathy R.

CONTENTS

10.1 INTRODUCTION

During the past decade, nanoparticles are being known for their use in marketable products, but their biomedical applications, including drug or gene delivery systems, biosensors, cancer treatment, and diagnostic tools, have been widely studied (Heera and Shanmugam, 2015). Nanoparticles are generally one-dimensional nanomaterial in the nanoscale range (Rallini and Kenny, 2017; Over, 2012). Nanoclay is one of the classifications of nanoparticles and is commonly used in polymer nanocomposites.

Nanocomposites are replacing the conventional type of micro-composites due to their improved properties. Due to lack of knowledge in these areas, there are a lot of challenges observed in their preparations. In earlier days the preparation of nanocomposite passed through more difficulties (Asmatulu *et al.*, 2015). Because of their low cost, polymeric nanocomposites have been widely used in industry. Since 1989, the Toyota Company has been used in the synthesis of nylon-6 clay hybrids and utilized in the manufacture of automobile bodies (Okada and Usuki, 2006). The miscellaneous applications of polymer nanocomposite are shown in Figure 10.1.

DOI: 10.1201/9781003218692-10

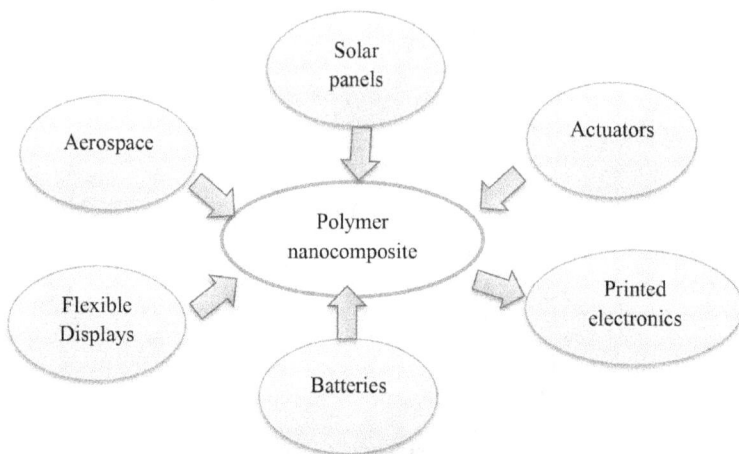

FIGURE 10.1   Recent advances in polymer nanocomposites.

Polymer foam can be categorized by its wall thickness, cell size and density, and comes in two main classes: closed-cell foam (cells are isolated) and open-cell foam (cells are connected). Polymer foams are observed in nature in the form of natural sponges, natural cork, bone and coral. Processing of polymer foams has received considerable attention motivated by the above materials. They are an excellent substitute for various functional materials due to their lightweight porous structure and they are used as barriers such as sound and thermal barriers, cushions, absorbents and shock absorbers. Also, polymeric foams suffer from poor surface quality, low mechanical strength, low dimensional and thermal stability (Klempner *et al.*, 2004). Moreover, enhanced physical and mechanical properties, fire resistance and the heat distortion temperature of polymer foams may be due to the presence of nanoparticles. A new class of materials with high strength, lightweight and multifunctional novel nanocomposite foams are developed which are based on the combination of functional nanoparticles and supercritical fluid foaming technology. In this article, we have been reviewing the recent progress in the area of polymeric foams.

As a matter of fact, a common classification of polymer foams and thus of polymer nanocomposite foams take into account their type of cellular structure: fully open-cell foams, interconnected or partially open-cell foams, and closed-cell foams. Most foaming processes lead to polymer nanocomposite foams with a distinctive closed-cell structure, albeit the addition of certain nanoparticles and foaming conditions may promote the partial opening of the cells, thus extending their possible range of applications. In general, closed-cell foams present a better balance of mechanical and insulation properties, hence being vastly used as thermal insulating and lightweight structural elements, while open-cell foams may be used for sound damping or as filters. Among closed-cell foams, a new sub-classification has appeared recently in order to consider closed-cell foams formed by an extremely high number of uniformly dispersed micrometric- or even sub-micrometric-sized cells: microcellular foams.

The main types of polymer foaming processes include
Continuous processes to prepare rigid PU foamed sheets

1. Extrusion (chemical or physical foaming)
2. Injection-molding (chemical foaming or physical foaming)

Batch foaming processes

1. Slabstock and mold processes to prepare flexible polyurethane (PU) foams
2. Rigid PU foams discontinuous processes
3. Compression-molding
4. High pressure gas dissolution
5. Expanded bead foams (Kumar and Nadella, 2004)

In spite of extensive research in the synthesis of polymer nanocomposites, some limitations persist that need further investigation. In some cases, inconsistent feedstock material, homogeneous dispersion, little awareness of the shape and size of filler material and unlike the nature of most filler material to polymers are the common problems face by particles-reinforced polymer composites. The material's composition, interfacial interaction or bonding and structural defects need to be analyzed in detail to confirm outstanding product performance (Wu *et al.,* 2020).

10.2   PROPERTIES

The use of additives as nucleating agents is one of the conventional methods to control the inducement of desired cell structures and foam morphology. Colton and Suh, 1987 have revealed that through the incorporation of a small amount of an additive in combination with a high saturation pressure of the blowing agent, a considerable increase in the pore cell population can be achieved (at a concentration level below its solubility limit). These effects are synergistic, in that they lower the intermolecular potential energy and interfacial tension and thereby increasing the number of small pore cells and reducing the activation energy barrier for nucleation. Nanoparticles have major effects on the stabilization of nucleated bubbles, especially at high temperatures. These studies provide essential insight into the role of heterogeneous nucleation in polymer foaming. Any characteristic aspects of composites could affect the final morphology of foamed nanocomposites such as the dispersion and distribution of particles and the filler content (Famili *et al.,* 2010). In comparison with conventional micrometer-sized filler particles, nanoparticles have greater specific surface areas. Therefore, at a low nominal particle concentration, a significantly higher effective particle concentration can be achieved. These factors could lead to improved nucleation efficiency (Tomasko *et al.,* 2003). The surface properties of nanoparticles include geometrical properties, such as the particle size, shape and size distribution. There are limited reports on the effects of the geometry of particles on efficiency of foaming. Shen *et al.,* 2005 compared the nucleation efficiency of single-walled carbon nanotubes (SWCNTs), nanoclays and carbon nanofibers. They verified that carbon nanofibers displayed substantially higher nucleation efficiency in comparison with SWCNTs and nanoclay. They specified that this was due to their good dispersion in the polymer matrix as well as the surface curvature and favorable wettability in the foaming process (Famili *et al.*, 2010).

Because of their unique multi-functionality, polymer foams are utilized in various applications. With the increase in the demand for lightweight structures, polymer foams with superior properties at low density have become popular. The properties of neat polymer foams are usually inferior compared to the properties of their solid counterparts due to the inclusion of air bubbles. Gibson and Ashby, 1982 developed one of the most commonly used models to quantitatively predict the mechanical properties of polymeric foams. Here, the polymer foams are supposed to be composed of unit cells with cell walls (for closed cell foams), the gas inside the cells and cell edges. All of these three components attribute to the mechanical properties of polymeric foams. When the gas pressure is atmospheric pressure, the influence of the gas compression is considered negligible. Acoustic damping efficiency is another distinguished property of polymer foams that can be influenced by nanofillers. Polymer foams are widely used as sound insulation materials because of the dissipation of sound energy as heat due to viscous friction between air friction inside cells and polymer chains. Providing thermal insulation is one of the major functions of polymer foams. Thermal insulation materials are extensively used in motor vehicles and buildings to diminish energy consumption. The effective thermal conductivity of polymer foams is due to heat flow in the solid polymer and the cell gas, convection processes and thermal radiation (Chen *et al.*, 2013). A schematic representation of the various properties of nanocomposites is demonstrated in Figure 10.2.

Through the fine characteristics and incorporation of efficient nanoparticles in matrices, polymer nanocomposite foams have gained greater interest (Kausar, 2018). CNT is one of the most commonly used nanocarbon to develop nanocomposite

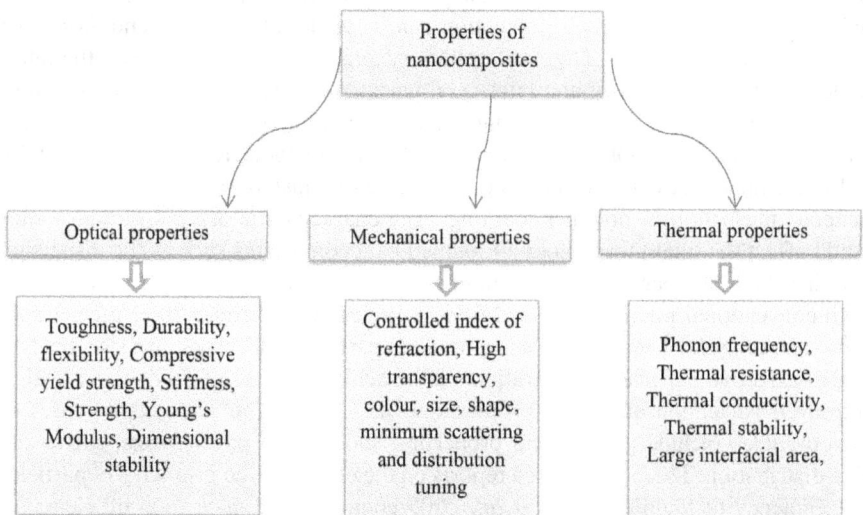

FIGURE 10.2   Various properties of the nanocomposite.

foams (Navidfar *et al.*, 2018). CNT foam is often termed CNT aerogel. CNT foam or aerogel is an electrically conducting and strong material. The recent technologies for electrodes, thermoelectric devices and sensors have been upgraded using CNT foam. CNT foams and their composite 3D materials have been developed using several approaches including freeze-drying, foaming agents, chemical vapor deposition (CVD) and chemical fusion methods. The CVD process may include a covalent junction between the nanotubes to improve the mechanical performance of foams (Shan *et al.*, 2013; Lin *et al.*, 2015).

10.3   APPLICATIONS OF NANOPARTICLES, NANOCOMPOSITES AND POLYMERIC NANOCOMPOSITE FOAMS

Polymeric materials have vast application from domestic to aerospace and are thus mostly studied and used materials in structural applications. The same enormous applications are anticipated from polymer matrix nanocomposites. Inorganic materials are generally used for strengthening polymer nanocomposites. Polymeric nanocomposite has good electrical conductivity and exceptional mechanical properties. So, it has an excellent choice for structural applications (Shenhar *et al.*, 2005). Recently, some researchers have attempted to produce hybrid polymer nanocomposites to improve the desired properties of polymer nanocomposites. The rapid manufacturing of gastrointestinal stents of polycaprolactone-polydioxanone (PCL-PDO) composite, varying their relative content was reported by Fathi *et al.*, 2020. *In vivo* and *ex vivo* investigations showed the materials' biocompatibility, and the material was found supportive of cell growth.

The production of sustainable and renewable storage resources in reservoirs for renewable energy such as the wind and the solar energy sector is urgently needed due to the growing energy crisis caused by the depletion of conventional fossil fuels (Cook *et al.*, 2010). The research importance of the Electrochemical Energy Storage Systems (EESS) with electrochemical capacitors (ECs), and fuel cells among diverse storage systems has been important such as thermal energy, compressed air energy storage (Saleh et al. 2020), hydro-pumped storage and flywheel storage (Manthiram *et al.*, 2008). The proclivity to substitute certain components with polymers is to decrease costs for dye-solar cells (DSCs) and maximize performance. For regular rigid and liquid electrolyte-based DSCs, 13% power conversion efficiency (PCEs) has already been accomplished (Mathew *et al.*, 2014). DSCs' key disadvantages particularly when used in portable electronics, are aspects such as rigidity and high weight, frangibility of the glass electrode, leakage and instability are the difficulties of liquid electrolytes (Tawfik *et al.*, 2020). Synthetic biodegradable polymer nanocomposites including bioactive ceramic phases are being increasingly considered for use as tissue engineering frameworks because of their enhanced biological, physical, and mechanical properties, as well as having the capability for modifying their structure and degradation rate to the specific requirement at the implant site. Usually, scaffolds should have appropriate mechanical properties such as elastic constants, with values in a similar range to those of the tissues at the site of implantation. A lot of research is focused on the progress of such porous bioactive and biodegradable nanocomposite frameworks for the repair and renewal of bone tissue. Among this group of materials, porous degradable polylactide, polyglycolide, or their co-polymers, are specifically relevant for bone tissue engineering, with additions of inorganic fibers or particles to impart bioactivity and improve mechanical properties (Christopher and Monika, 2008).

Shape memory polymer (SMP) foams are studied as hemostatic biomaterials. SMPs are defined as smart stimuli-responsive materials that are synthesized in a permanent, primary shape, triggered using an external stimulus, such as light or heat, and fixed or strained into a temporary, secondary shape that is recollected upon removal of the external stimulus. After re-exposure to the stimulus, SMPs recover their primary shape. Heat is used as an external stimulus in some polyurethane SMP foam systems, and shape change is designed around the polymer's glass-transition temperature (Tg). These polyurethane SMP foams are biocompatible and capable of encouraging rapid blood clotting due to their high surface area and thrombogenic surface chemistry (Monroe *et al.*, 2018; Vakil *et al.*, 2021).

SMPs have shown increasing attention in the synthesis of nanocomposites for functional applications. These materials can recover their original shape after deformation due to any external stimuli. SMPs can be reinforced with nanoparticles to obtain shape memory nanocomposites with controlled actuation, improved mechanical properties, and recovery mechanisms. SMP nanocomposites have considerable potential for their applications in 4D printing, electronics, biomedical and aerospace. Xia *et al.*, 2021 reported on these materials and their applications. Recent researches also aim to accept the bio-inspired design to the medical field to synthesize soft materials (rubber, leather and polymers) that can be used for the manufacture of delicate organs and tissues. In this respect, additive manufacturing would be a potential candidate for integration to realize advanced materials. One of the main advantages of employing additive manufacturing in the medical field is its ability to therapeutic approaches and to produce patient-specific devices (Rashid *et al.*, 2021).

Ajayan *et al.*, developed a controlled method of manufacturing isolated polymer nanotube-composite films that can be used to form nanosensors, which contains at least one conductive channel including an array of aligned CNTs embedded in a matrix like poly-dimethylsiloxane. These materials can be used to regulate a real-time physical condition of a material and may be used for the identification and detection of poisonous gases such as flammable gases, solvent vapors, chemical warfare agents, etc., touch sensors for interaction between operators and machines (Gentle and Baney, 1992). Polymer nanocomposites as microwave absorbers are receiving much attention. Nguyen and Diaz have reported a method to synthesize polypyrrole nanocomposites containing titanium dioxide, tungsten oxide, tin oxide and iron oxides (α and β). The magnetic properties of the pyrrole containing a dispersion of nanoparticle metal oxides were polymerized *in situ* reported. The promising application of nanocomposites in body armor was also reported (Lee *et al.*, 2002).

Polymer nanocomposites have been proven to be the most promising and preferred nanomaterials compared to other functional nanocomposites. Because of their remarkable performances to advance engineering applications such as outstanding mechanical, electrical and thermal properties, they have attracted significant interest from researchers in the various fields of science and technology. Polymer nanocomposites have been systematically studied since they were discovered as a nanomaterial. Their extraordinary properties have made them encouraging in many areas, including defense equipment, automotive, sports goods manufacturing, energy field, sensors and supercapacitors. The dispersal of reinforced nanomaterials has a key role in attaining desired physio-chemical properties of polymer nanocomposites. On the contrary, many processing approaches used to synthesize these polymer nanocomposites are not viable economically. Layer-by-layer assembly techniques, electrospinning techniques and solvent processing techniques still result in perfect homogeneous dispersions of reinforced materials in the polymer are not so profitable (Shukla and Saxena, 2021).

Polymer nanocomposite has improved mechanical properties, barriers, impact, and heat resistance than conventional composites. So, the development of polymer nanocomposite having properties like recyclability, lightweight and biodegradability is a challenge. Such nanocomposites can be extensively used for the manufacturing of automobile body parts. Industries are concerned mainly with the following aspects like aesthetics, recyclability, reduction of weight and performance improvements.

Recently, there has been a significant impulse in the development of polymer nanocomposite foams, extending the idea of polymer nanocomposites to the field of cellular materials. The progress in new advanced foaming technologies have permitted the generation of new foams with nano, sub-micro and microcellular structures. This extended the applications of more traditional foams in terms of damping, weight reduction, and thermal or acoustic insulation to novel possibilities, like electromagnetic interference (EMI) shielding. Ma *et al.*, have concentrated on both theoretical and experimental methods to evaluate the flame spreading and mechanical characteristics of rigid PU foams. This can be used as insulation materials in building elevation, clarifying some features of downward flame spreading over PU foams, with direct implications in the design of safe vertical facades for buildings.

Chen *et al.*, have examined the possibility of improving the flame retardancy and thermal stability of rigid PU foams using functionalized graphene oxide (fGO). The matrix is in combination with the well-known intumescent system formed by expandable graphite (EG) and dimethyl methyl phosphonate (DMMP). Here, experimentally demonstrated the improvement of the thermal stability and a decrease in the flammability of PU foams even by adding extremely low amounts of fGO, which acted to strengthen the intumescent char layer formed during burning. Similarly, Xi and co-workers have used ternary flame retardant systems formed by aluminum hydroxide (ATH), EG and [bis(2-hydroxyethyl)amino]-methyl-phosphonic acid dimethyl ester (BH) to enhance the flame retardancy of rigid PU foams.

In recent times, flame-retardant graphene foams (GFs) that are ultralight and also show compressible characteristics were reported (Hu *et al.*, 2016). Due to their flexible chemical structures, polymer foams are also lightweight and used in various applications (Chen *et al.*, 2013). Unfortunately, polymer foams are susceptible to catching fire. Over the past several decades, various types of flame retardants, including phosphorus- and halogen-based agents (Benin *et al.*, 2012) and other flame-resistant materials (Ruan *et al.*, 2014) have been successfully incorporated into a wide range of polymer foams to minimize flammability. Xu *et al.*, 2017 reported the fabrication of superior flame-retardant polyimide (PI) foams that are also of high strength and lightweight through the incorporation of red phosphorus hybridized graphene.

Thermoplastic polyurethane (TPU) foam has been known as positive triboelectric material as per the static electric series provided (Henniker, 1962). The triboelectric effect is a type of contact electrification in which certain materials become electrically charged after they are separated from a different material with which they are in contact. TPU foam is a potential triboelectric material and well-accepted artificial skin for medical diagnostics (Morimoto *et al.*, 216). Exclusive research opportunities are done there for copying human skin and development of biocompatible bi-layer artificial skin for tactile sensing performance. Li *et al.*, 2018 reported the relative foaming performance of CO_2 over N_2 to design a TPU foam-based sliding mode triboelectric nanogenerator as prototype bi-layer artificial skin having flexibility and adequate softness, which can potentially distinguish different objects due to contact electrification.

GFs are versatile nanoplatforms for biomedical applications due to their excellent mechanical, physical and chemical properties. However, the inflexibility and brittleness of pristine GF (pGF) are some of the important factors limiting their extensive application. Main challenges in graphene-reinforced polymer composites (GRPC) which have encouraged the need for GF filler in polymers to include high inter-sheet junction contact resistance, weak interfacial interaction between graphene and polymer matrices and non-uniform distribution of graphene sheets within the polymer. The electronic industry can use the evolving GF-based polymer matrix composites in super-capacitors, flexible electronics, electronic devices, electrochemical biosensors, strain sensors and EMI shielding materials. GFs framework in polymers has resulted in improved interaction between the GFs filler and polymer matrices. The scope of knowledge about GF-based polymer matrix composites is yet to be discovered and thus can drive the research boundary of GFs-based composites (Idowu *et al.*, 2018).

Li *et al.*, 2011 have reported superhydrophobic microporous foams of conjugated polymers. The foams have revealed a high level of selectivity for oils. Through mechanical squeezing, solvent-washing method, or burning process, superhydrophobic three-dimensional foams can remove oils or organic pollutants. Superhydrophobic CNT-based foams have gained success for the separation or adsorption of oils and organic solvents. Kabiri *et al.*, 2014 reported polymer and CNT network foam structures which possess enhanced porosity, robustness, oleophilic properties and hydrophobicity (Kausar, 2019).

PI-based foam materials are advanced functional materials and are increasingly used as key materials for vibration, thermal insulation and noise reduction in aviation, military hardware, transportation, aerospace and other high-tech areas (Flores-Bonano *et al.*, 2019). In recent years, conductive polyurethane composite foams have attracted more attention due to their thermal insulation properties and low density (Ma et al. 2018). Furthermore, these materials are particularly stimulating for their efficiency in defending against EMI, which permits their use in electronic devices where electromagnetic emissions can interfere with the operation of neighboring electronic equipment (Gosh *et al.*, 2018).

One of the most trending improvements is in the use of CNT-based polymer foams as structural scaffolds for tissue engineering, as it has been shown that a proper design of an open-cell porous structure that may serve as supporting material for cell implantation and growth (scaffold) results crucial in guaranteeing both in vitro as well as *ex vivo* living tissue regeneration (Velasco, J. I., and M. Antunes). Specific attention has been given to the addition of CNTs to PU-based foaming systems as, on the one hand, CNTs may enhance some properties and functionalities of the structural scaffolds (Harrison and Attala, 2007) such as mechanical properties or osteo conductivity and mineralization potential, and, secondly, being PU an initially liquid system, it facilitates the incorporation and dispersion of the nanotubes.

10.4   CONCLUSIONS

The latest developments made in the area of material nanoparticles reinforced polymer nanocomposite foams, their properties and some important applications have been reviewed. A variety of nanoparticles shows great potential as strengthening in polymer nanocomposite foams for improvement in diverse applications. The necessity for bulk production of polymer nanocomposite foams becomes the existing challenge in concerns with the economical processing techniques. Different techniques for preparing exfoliated polymer nanocomposite foams with required flammability, structural and other properties must be recognized. Bulk production of polymer nanocomposites and nanocomposite foams depends on consistent and sensible synthetic processing methods. This chapter reviews some of the recent and novel nanocomposite foams and their selected applications, but to move these materials into commercial products, several challenges must be overcome. With the emerging developments in the field of nanotechnology, novel nanoparticles will be discovered or synthesized, which may find potential applications in polymer nanocomposite foams. Generally, polymer nanocomposite foams offer better performances compared to conventional nanocomposites and these are the future materials of choice in all divisions of medicine, agriculture, engineering and energy. Thus it can be concluded that although polymer nanocomposite foams possess wide potential applications in biomedicine, more detailed studies are essential to obtain a higher understanding of the various interesting and promising technology. Therefore, to obtain this understanding new analytical tools are needed to accurately and reliably determine the cytotoxic studies of polymer nanocomposite foams. A multidisciplinary team of material scientists, molecular biologists, and toxicologists is also needed to guarantee that all aspects of nanotoxicity are considered. Such collaboration is a key to the safe design and development of nanocomposites because it makes it possible to understand the interactions of bioactive polymer nanocomposite foams.

REFERENCES

Asmatulu, R., Khan, W.S., Reddy, R.J., Ceylan, M. 2015 Synthesis and analysis of injection-molded nanocomposites of recycled high-density polyethylene incorporated with graphene nanoflakes, *Polym. Compos.* 36(9):1565–1573

Benin, V., Durganala, S., Morgan, A.B. 2012 Synthesis and flame retardant testing of new boronated and phosphonated aromatic compounds. *J. Mater. Chem.* 22: 1180–1190.

Chen, L.M., Rende, D., Schadler, L.S., Ozisik, R. 2013 Polymer nanocomposite foams. *J. Mater. Chem. A,* 1:3837–3850.

Colton, J.S., Suh, N.P. 1987 Nucleation of microcellular foam: Theory and practice. *Polym. Eng. Sci.*, 27:493–499.

Cook, T.R., Dogutan, D.K., Reece, S.Y., Surendranath, Y., Teets, T.S., Nocera, D.G. 2010 Solar energy supply and storage for the legacy and nonlegacy worlds, *Chem. Rev.*, 110:6474–6502.

Famili, M.H.N., Janani, H., Enayati, M.S. 2010 Foaming of a polymer-nanoparticle system: Effect of the particle properties. *J. Appl. Polym. Sci.*, 119(5):2847–2856.

Fathi, P., Capron, G., Tripathi, I., Misra, S., Ostadhossein, F., Selmic, L., Rowitz, B., Pan, D. 2020 Computed tomography-guided additive manufacturing of Personalized Absorbable Gastrointestinal Stents for intestinal fistulae and perforations. *Biomaterials* 228:119542

Flores-Bonano, S., Vargas-Martinez, J., Suarez, O.M., SilvaAraya, W. 2019 Tortuosity index based on dynamic mechanical properties of polyimide foam for aerospace applications. *Materials* 12 (11):1851.

Gentle, T.E., Baney, R.H. 1992 Silica/silicone nanocomposite films: A new concept in corrosion protection. *Proc. Mater. Res. Soc. Symp.*, 274:115–19.

Ghosh, S., Ganguly, S., Remanan, S., Mondal, S., Jana, S., Maji, P.K., Singha, N., Das, N.C. 2018. *J. Mater. Sci. Mater. Electron.*, 29:10177.

Gibson, I.J., Ashby, M.F. 1982 The mechanics of three-dimensional cellular materials. *Proc. R. Soc. Lond.*, A382:43–59.

Harrison, B.S., Attala, A. 2007 Carbon nanotube applications for tissue engineering. *Biomaterials*, 28:344–353.

Heera, P., Shanmugam, S. 2015 Nanoparticle characterization and application: an overview. *Int. J. Curr. Microbiol. App. Sci.*, 4 (8):379–386.

Henniker, J. 1962 Triboelectricity in polymers. *Nature*, 196:474.

Hu, C., Xue, J., Dong, L., Jiang, Y., Wang, X., Qu, L., Dai, L. 2016 Scalable preparation of multifunctional fire-retardant ultralight graphene foams. *ACS Nano*, 10:1325–1332.

Idowu, A., Boesl, B., Agarwal, A. 2018 3D graphene foam-reinforced polymer composites—A review. *Carbon*, 135:52–71.

Kabiri, S., Tran, D.N., Altalhi, T., Losic, D. 2014 Outstanding adsorption performance of graphene–carbon nanotube aerogels for continuous oil removal. *Carbon*, 80:523–533.

Kausar, A. 2018 Polyurethane composite foams in high-performance applications: A review. *Polym. Plast. Technol. Eng.*, 57:346–369.

Kausar, A. 2019. Advances in polymer-anchored carbon nanotube foam: a review. *Poly. Plast. Technol. Mater.*, 1–14. https://doi.org/10.1080/25740881.2019.1599945

Klempner, D., Sendijarevic, V., Aseeva, R.M. 2004 *Handbook of polymeric foams and foam technology*, Hanser Gardener Publications, Cincinnati, 2nd edn, 2004

Kumar, V., Nadella, V. 2004 *Microcellular foams. In Handbook of polymer foams*, ed. N. Mills, 243–267. Rapra Technology Limited, Shawbury.

Lee, Y.S., Wetzel, E.D., Egres Jr., R.G., Wagner, N.J. 2002 Advanced body armor utilizing shear thickening fluids. *In Proceedings of the 23rd Army Science Conference.*

Li, A., Sun, H.X., Tan, D.Z., Fan, W.J., Wen, S.H., Qing, X.J., Li, G.X., Li, S.Y., Deng, W.Q. 2011 Superhydrophobic conjugated microporous polymers for separation and adsorption. *Energy Environ. Sci.*, 4:2062–2065.

Li, H., Sinha, T.K., Oh, J.S., Kim, J.K. 2018 Soft and flexible bilayer thermoplastic polyurethane foam for development of bioinspired artificial skin. *ACS Appl. Mater. Interfaces*, 10(16):14008–14016.

Lin, Z., Gui, X., Gan, Q., Chen, W., Cheng, X., Liu, M., Zhu, Y., Yang, Y., Cao, A., Tang, Z. 2015 In-situ welding carbon nanotubes into a porous solid with super-high compressive strength and fatigue resistance. *Sci. Rep.*, 5:11336.

Ma, X., Tu, R., Cheng, X., Zhu, S., Ma, J., Fang, T., 2018 Experimental study of thermal behavior of insulation material rigid polyurethane in parallel, symmetric, and adjacent building façade constructions. *Polymers,* 10:1104.

Manthiram, A., Vadivel, M. 2008 Nanostructured electrode materials for electrochemical energy storage and conversion. *Energy Environ. Sci.,* 1:621–638.

Mathew, S., Yella, A., Gao, P., Humphry-Baker, R., Curchod, B.F.E., Ashari Astani, N. 2014 Dye sensitized solar cells with 13% efficiency achieved through the molecular engineering of porphyrin sensitizers. *Nat. Chem.,* 6:242–247

Monroe, M.B.B., Easley, A.D., Grant, K., Fletcher, G.K., Boyer, C., Maitland, D.J. 2018 Multifunctional shape memory polymer foams with bio-inspired antimicrobials. *Chem. Phys. Chem.,* 19:1999–2008.

Morimoto, N., Kuro, A., Yamauchi, T., Horiuchi, A., Kakudo, N., Sakamoto, M., Kusumoto K. 2016 Combined use of fenestrated-type artificial dermis and topical negative pressure wound therapy for the venous leg ulcer of a rheumatoid arthritis patient. *Int. Wound J.,* 13:137–140.

Navidfar, A., Sancak, A., Yildirim, K.B., Trabzon, L. 2018 A study on polyurethane hybrid nanocomposite foams reinforced with multiwalled carbon nanotubes and silica nanoparticles. *Polym. Plast. Technol. Eng.,* 57:1463–1473.

Nguyen, T., Diaz, A.F. 1994 Novel method for the preparation of magnetic nanoparticles in a polypyrrole powder. *Adv. Mater.,* 6(11):858–60.

Okada, A., Usuki, A. 2006 Twenty years of polymer-clay nanocomposites. *Macromol. Mater. Eng.,* 291:1449–1476.

Over, H. 2012 Surface chemistry of ruthenium dioxide in heterogeneous catalysis and electrocatalysis: From fundamental to applied research. *Chem. Rev.,* 112(6):3356–3426.

Rallini, M., Kenny, J.M. 2017 3-Nanofllers in polymers. *Modification Polym. Prop.,* 47–86. https://doi.org/10.1016/B978-0-323-44353-1.00003-8.

Rashid, A.A., Khan, S.A., Al-Ghamdi, S.G., Koc, M. 2021 Additive manufacturing of polymer nanocomposites: Needs and challenges in materials, processes, and applications, *J. Mat. Res. Tech.,* 14:910–941.

Ruan, C., Ai, K., Li, X., Lu, L. 2014 A superhydrophobic sponge with excellent absorbency and flame retardancy. *Angew. Chem., Int. Ed.,* 53:5556–5560.

Saleh, T.A., Shetti, N.P., Shanbhag, M.M., Reddy, K.R., Aminabhavi, T.M. 2020 Recent trends in functionalized nanoparticles loaded polymeric composites: An energy application. *Mater. Sci. Energy Technol.,* 3:515–525.

Shan, C., Zhao, W., Lu, X.L., O'Brien, D.J., Li, Y., Cao, Z., Elias, A.L., Cruz-Silva, R., Terrones, M., Wei, B. 2013 Three-dimensional nitrogen-doped multiwall carbon nanotube sponges with tunable properties. *Nano Lett.,* 13:5514–5520.

Shen, J., Zeng, C., Lee, L.J. 2005 Synthesis of polystyrene–carbon nanofibers, nanocomposite foams. *Polymer,* 46: 5218–5224.

Shenhar, R., Norsten, T.B., Rotello, V.M. 2005 Polymer-mediated nanoparticle assembly: Structural control and applications. *Adv. Mater.,* 17(6):657–669.

Shukla, P., Saxena, P. 2021 Polymer Nanocomposites in sensor applications: A review on present trends and future scope. *Chin. J. Polym. Sci.,* https://doi.org/10.1007/s10118-021-2553-8

Tomasko, D.L., Han, X., Liu, D., Gao, W. 2003 Supercritical fluid applications in polymer nanocomposites. *Curr. Opin. Solid State Mater. Sci.,* 7:407–412.

Vakil, A.U., Petryk, N.M., Shepherd, E., Beaman, H.T., Ganesh, P.S., Dong, K.S., Monroe, M.B.B. 2021 Shape memory polymer foams with tunable degradation profiles. *ACS Appl. Bio Mater.,* https://doi.org/10.1021/acsabm.1c00516.

Velasco, J.I., Antunes, M. PLA-based foams and related porous materials for tissue engineering. *In Encyclopaedia of Biomedical Polymers and Polymeric Biomaterials,* ed. M. Mishra (in press).

Wu, H., Fahy, W.P., Kim, S., Kim, H., Zhao, N., Pilato, L., Kafi, A., Bateman, S., Koo, J.H.
    2020 Recent developments in polymers/polymer nanocomposites for additive manufac-
    turing. *Prog. Mater. Sci.,* 111:100638.
Xia, Y., He, Y., Zhang, F., Liu, Y., Leng, J. 2021 A review of shape memory polymers and
    composites: mechanisms, materials, and applications. *Adv Mater.,* 33:1–33.
Xu, L., Xiao, L., Jia, P., Goossens, K., Liu, P., Li, H., Cheng, C., Huang, Y., Bielawski, C.W.,
    Geng, J. 2017 Lightweight and ultrastrong polymer foams with unusually superior flame
    retardancy, *ACS Appl. Mater. Interfaces*, 9:26392–26399.

11 Innovations in Polymeric Foams and New Application Opportunities Including Energy and Energy Devices

Nizam P. A. and Sabu Thomas

CONTENTS

11.1 INTRODUCTION

Polymer foams are gaining popularity in advanced and developing industries that require more efficient materials to meet challenging technical criteria such as improved property and lower energy consumption. They continue to expand due to their exceptional properties such as strength-to-weight ratio, lightweight, thermal, and insulating properties, good cushioning, and energy absorption properties. It was designed to save costs and minimize the quantity of plastic polystyrene required in industrial operations, and since it has made significant progress. These foam materials are primarily used in the construction, automotive, aerospace, packaging, electronics, and medical sectors.

In recent times due to the rapid expansion in technology, there has been a massive increase in innovations in almost all sectors, which includes, aerospace, food

DOI: 10.1201/9781003218692-11

industry, and electronics. In each sector, there arises lots of research to employ various materials for their applications. We're talking about the electronic industry, where new technologies are being launched daily, for example, SCs, wearable electronics, and so on. Polymers and nanomaterials are among the most investigated materials owing to their superior physical and chemical characteristics. Polymer foams are used in a variety of applications, including SCs, EMI shielding, EEG, energy absorption, and so on.

SCs are novel energy storage devices with characteristics such as high capacitance, high power density, and long cycle life. Higher stability, durability, and catalytic activity, as well as lower cost, are critical features of materials utilized in SCs. EMI shielding is another sector in electronics where materials are investigated to reduce interference either by reflection or absorption mechanism. Human EEG provides a handy, but sometimes opaque, 'window on the mind,' allowing one to observe electrical activity at the brain surface. Polymers are employed in these sectors by improving the properties of the same, thanks to the tailorable characteristics of the polymers. Polymer foams due to their porous structure as well as surface areas enhance their use in these applications.

This chapter emphasis on sustainable developments in foam technology where sustainable and environmentally friendly monomers are employed for their manufacture. A study on biodegradable foams is also at a fast pace. Furthermore, the central theme of the chapter is the use of polymer foams in energy sectors, which encompasses the applications as well as others such as acoustic application and energy absorption.

11.2 ADVANCES IN SUSTAINABLE POLYMER FOAMS

Polyurethane foams (PUFs) are the most significant family of polymeric foams, due to their thermal conductivity, low density, and unique mechanical properties, which make them great thermal and sound insulators, as well as structural and comfort materials. Polyurethane (PUs) are polymers generated when the OH (hydroxyl) groups of a polyol react with the NCO (isocyanate functional group) groups of an isocyanate. The technique for producing PUFs is so well-established when compared to other polymeric foams, for which they account for 50% of worldwide total foams. PUFs are used in home and industrial applications, as well as space travel and medicine, due to their durability and adaptability. A vast characteristic feature can be achieved by changes in composition which enhances their adaptability in various applications.

Sustainable materials are the key focuses of the current generation which is being investigated in numerous industries to replace renewable resources. Many renewable sources are being employed in the manufacture of PU foams and other PU products as a result of rising environmental concerns and diminishing petroleum supplies. To manufacture environmental-friendly PU goods, polyols, a basic ingredient for PU foam, have been generated from several renewable sources such as vegetable oils, sugarcane bagasse, and pine wood. Polyols were made from soybean oil via epoxidation and hydroxylation. The sustainable PU foams were prepared using a free-rise process and then compared to petroleum-based PU foams in cars and bedding cushions. The inherent molecular chain enhances a higher Tg, inferior cryogenic characteristics, and lower thermal degradation temperatures. The good mechanical characteristics of these bio foams offer up possibilities for employing them in applications if the poor attributes are addressed (Gu, Konar and Sain, 2012).

The renewable sources of polyol can be divided into two depending upon the number of hydroxyl groups present in low and high hydroxyl groups. Vegetable oils are excellent precursors but lack enough hydroxyl groups which makes it unacceptable to fabricate foams. In this case, hydroxyl groups can be incorporated into their chains. In the case of Castrol oil, two processes are employed to increase the hydroxyl groups which are transesterification using their ester functional group and alkoxylation of castor. Transesterification can be done using polyfunctional alcohols such as glycerols, pentaerythritol, triethanolamine, and trimethylolpropane which improves the hydroxyl moieties. Rapeseed and soybean oil utilizes an epoxidation process followed by transesterification (Zhang *et al.*, 2013).

Flexible PU foams were fabricated and comparative studies were carried out using polyol based on soybean, cross-linked polyols, and styrene-acrylonitrile copolymer-filled polyols. The result exhibited an improved compression modulus for all the foams with the best result from 30% soybean-based foam. Morphology changes in these foams are well explained by the increase in compression modulus (Miao *et al.*, 2012).

Corncobs, pinewood, and eucalyptus are some of the other renewable sources of polyols utilized to make environmentally friendly PU goods. They are rich in hydroxyl groups as their components include lignocellulosic materials. Lignin-based foams were fabricated by substituting various quantities of petroleum polyols. The foams have bigger cell sizes and low foam density as these foams are embedded with pulp fibers. They exhibited identical properties as that synthetic foams in density, thermal conductivity, compressive strength, and elasticity. These foams showed better degradable properties when compared to synthetic foams (Cateto *et al.*, 2014).

Most of the polymers are processed from fossil fuels and mostly they end up as pollutant waste. Researchers are rapidly trying to develop green and biodegradable polymers for future generation. Poly (lactide acid) or polylactide (PLA), is one of the best substitutes for synthetic polymers. They are made from sugarcane and cornstarch. They are aliphatic thermoplastic polyester synthesized from biodegradable and renewable resources. These foams are already in research and they are an alternative to many petroleum-based polymer foams. Polystyrene can be replaced by PLA in applications such as cushioning, packaging, and thermal and acoustic insulations. PLA has very low processing and material cost and often has good mechanical properties and biocompatibility. For these reasons, PLA can be employed in biomedical applications which include scaffolds and tissue engineering (TE). However, the low melt strength and delayed crystallization kinetics are some of the flaws of PLA. This restricts fabricating low-density foams with consistent cell shapes (Nofar and Park, 2014).

A PLA foam with thermal conductivity of 30 mW/m-K was developed using a supercritical CO_2 (sCO_2) foaming method. The high expansion ratio and excellent infrared (IR) block capabilities of PLA foam substantially aided in achieving this result. Unlike PS foams, which frequently contain non-biodegradable carbon particles to block IR heat radiation, the PLA foam's inherent IR-absorbing property, which was mediated by the ester group in the PLA molecular chain, boosted its environmental effect. Overall, PLA foams, which are created utilizing the non-toxic supercritical CO_2 foaming process, are a viable alternative to PS foams (Gong *et al.*, 2018). Despite their biodegradability, biocompatibility, and simplicity of production with desired morphologies, PLA and PLA-based foams have been widely studied as scaffolds for TE applications. PLA meets some of the criteria for TE applications, but it falls short on others, including bioactive activity and acceptable mechanical properties for some applications (Shah Mohammadi, Bureau and Nazhat, 2014). As a result, current research has focused on the inclusion of bioactive ceramics or glasses, as well as bifactors (such as proteins, genes, and growth hormones) into PLA scaffolds. used $scCO_2$ and an alginate entrapment approach to creating composite scaffolds of alginate and PLA that allowed for a variety of degradation rates and bio factor release. The encapsulation of vascular endothelial growth factor (VEGF) and bone morphogenetic protein (BMP) in alginate fibers and PDLLA resulted in a faster release of VEGF and a slower release of BMP, both of which are important in bone regeneration (Ginty *et al.*, 2008).

11.3 ADVANCES IN THERMOPLASTIC FOAMS AND PROCESSING TECHNIQUES

Thermoplastic foams are used in a variety of applications due to their excellent properties which counteract their solid type. The advantages include lightweight, material savings, enhanced toughness, accuracy, better fatigue life, better insulation properties, and acoustic properties. Batch foaming, bead foaming, compression molding, rotational molding, extrusion foaming, and foam injection molding (FIM) are all methods for making thermoplastic foams. Extrusion and injection molding methods have received a lot of attention because of their great products in the plastics industry.

FIM is a cutting-edge manufacturing method that evolved from the traditional injection molding process. Gas is injected into the molten polymer in FIM using either a physical blowing agent (PBA) or a chemical blowing agent (CBA), depending on the gas-producing method. For homogeneous dispersion of a blowing agent, gas and molten polymer are thoroughly blended. Following that, a polymer and gas combination is injected into a cavity, which is then shaped after it has cooled. Weight reduction, material cost savings, faster cycle times, higher-dimensional stabilities, and the lack of sink marks on surfaces are all advantages of the FIM technique. However, as compared to solid equivalents, the use of FIM technology degrades the mechanical characteristics of foam products. Furthermore, due to the fountain flow pattern caused by premature cell nucleation and strong cell growth near the flow front, FIM might leave swirl traces on the surface. Two types of pressure-based molding techniques were developed to address the issue, low and high-pressure FIM.

Low-pressure FIM is distinguished by two characteristics: a relative cavity pressure ranges from 0.5 to 10 MPa and a lower injection shot size of 65–80 vol% of the whole shot volume. The injection pressure is reduced, resulting in a lesser tonnage in the injection molding machine because of the smaller shot size. Furthermore, the use of low-pressure FIM might save a significant amount of money on tooling. They are frequently used in applications with big and thick-walled items because it permits a considerable amount of the cavity volume to be filled by expansion. Other benefits include reduced residual stress, improved dimension stability, and lower material costs. As a result, items that provide simple purposes, such as packaging materials and pallets, are common uses. The high-pressure process principle includes rapid complete filling of the cavity, core foaming separation, and formation of the solid skin layer. As the cavity is filled with the full shot size injection initially, and a tiny degree of volume becomes accessible during the cooling phase owing to polymer shrinkage, the volume expansion for high-pressure FIM is equal to the amount of shrinkage. The homogeneity of the foam structure is the key benefit of high-pressure FIM. Furthermore, high-pressure FIM can enhance surface properties by reducing the impact of flow swirls. However, the degree of volume expansion is quite limited, making material cost savings a challenge.

Gas counter pressure (GCP) technology was developed as a strategy to improve the surface properties of traditional FIM products. GCP is an additional technique that pressurizes a cavity before injecting the polymer/gas matrix. They restrict aggressive cell development at the leading edge of the flow front preventing surface swirl caused by fountain flow. To efficiently deploy GCP technology, a pressure-tight cavity design is necessary to reduce pressure loss throughout the GCP process, and the cavity is typically pressurized using N2 gas. After pressurizing the cavity to the required pressure, a predefined volume (i.e., shot size) of the compressed polymer/gas matrix is injected into the cavity. Controlled venting is necessary during this injection phase to maintain a steady counter pressure as the unfilled cavity capacity is lowered. After the injection is completed and an adequate thickness of solid, smooth skin layer has developed, the counter-pressure is entirely removed to induce foaming from the molten core (Jung, 2013).

An extruder usually consists of two counter-revolving screws within a barrel and an opening at one end to push the materials through the orifice. These screws and the clearance between the screw and barrel incorporate pressure into the sample to process well. This mechanical and thermal treatment enhances the processability whilst the end texture is governed by the die. Thermoplastic foaming is done using this method for many applications. The use of a CBA is commonly used to create foams (CBA). Blowing agents are low molecular weight substances, usually combined with polymer matrices. At some temperature, these materials decompose and a gas is formed. This basic technique does not provide adequate control of porosity, and the cellular structures of the products are frequently non-uniform. This also suggests that residues will be present in the final product, necessitating the insertion of a second step to remove them. PBAs, are widely employed to address these issues. Porosities are produced in the final product when these materials evaporate due to changes in temperature and pressure. Carbon dioxide is better former than inert gases as they are easy to handle and has better interactions with other polymers.

An intriguing material with characteristics between gases and liquids is a supercritical fluid. They possess temperature and pressure values high than their critical values. One example is carbon dioxide which is commonly used due to its properties such as non-flammability, non-toxicity, inertness, and the ease of attaining supercritical levels. Supercritical CO_2 is used in various polymerization process which includes foaming, polymer composites, and manufacturing of particles. Their unique characteristics of liquid-like density and gas-type viscosity enhance their use as a good plasticizer thus improving the rheological properties of material within the extruder. They expand the polymers within the barrel to create end products. Mechanical loads will be limited, and operating temperatures will be reduced, as viscosity decreases. It will also enable the handling of molecules with low thermal stability. This volume expansion due to pressure-induced will, bee recovered upon the air pressure. These blowing agents puff the material at the die with a temperature of less than 100°C which decouples the role of water as a plasticizer and as a blowing agent during food product extrusion. Output is improved with reduced moisture content (Sauceau *et al.*, 2011). The implementation of this CO_2 in extrusion has expanded the applications of extrusion operations. Almost all thermoplastics can be employed through this process, which opens up new possibilities in polymer industries.

11.4 APPLICATION OF POLYMER FOAMS IN SCS

SCs are energy storage devices that combine the advantages of rechargeable batteries with regular capacitors, offering both high energy storage and great charge/discharge efficiency. Their unique characteristics such as life cycle, maintenance, electrochemical performance, operating temperature, and environmental protection have augmented their potential to apply in electronics. With the growing popularity of elastic electronics and wearable devices, SCs are being expected to meet ever-higher standards, such as flexibility, stretchability, compressibility, and so forth. Compressible SCs (CSCs) have additional problems in design and manufacturing than other flexible wearables SCs because they must maintain outstanding capacitance performance despite undergoing considerable external strain and shape deformation, such as folding, bending, twisting, and compression. The primary issue in developing CSCs is to use electrodes that are structurally strong and deformation-tolerant while still having a high energy density. Sun et al. developed a flexible and compressible polypyrrole-coated melamine electrode foam, using an interfacial polymerization technique. These electrodes displayed strong cyclic stability and specific area capacitance, with 80% capacitance even after 3,000 cycles. A symmetrical SC was constructed with high energy density and outstanding cyclic stability after the current density is increased tenfold. The research demonstrated a deformation enhances wearable energy storage devices (Sun *et al.*, 2019). Figure 11.1 depicts the various concentration of MF/PPy composite and their subsequent increase in the intensity of light with increasing concentration.

For electrical and energy applications, increased nanoscale surface area and stretchability are major criteria. A generic technique for forming a customized wrinkling pattern on surfaces inside a 3D polydimethylsiloxane (PDMS) polymeric foam was fabricated. The wrinkled 3D foam enhances the specific surface area and stretchability by over 60 and 75%, respectively, after 1 h of plasma treatment and 40% pre-stretching. SCs are made by coating a conductive substance over the wrinkled foam to demonstrate its application with increased performance. The resultant

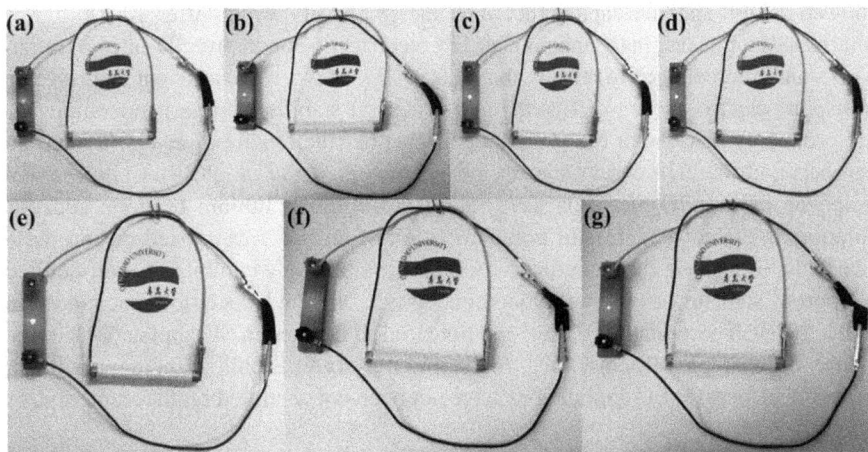

FIGURE 11.1 Various concentrations of MF/PPy are used as a part of the wire to light blue LED. Reprinted from the reference (Sun *et al.*, 2019).

SCs have an improved storage capacity of 8.3 times that of conventional SCs and can withstand stretching up to 50% (Jang *et al.*, 2021).

Aerogels are porous materials with variable physicochemical properties which include high porosity, low density, controllable surface chemistry, and large surface area. The potential of this porous material for a variety of applications has been thoroughly studied thanks to recent breakthroughs in the synthesis techniques of different aerogels. Different types of aerogels have been fabricated with advances in aerogels and trying technologies. Inorganic aerogels include TiO_2, SiO_2, ZrO_2, and $Al2O_3$. Organic aerogels mostly consist of resorcinol-formaldehyde PU and poly-styrene polyimide. Conductive aerogels are explored widely as they comprise high porosity, low density, and high surface area with electrically conducting materials embedded. Pectin and aniline-based mechanically strong conducting aerogels were fabricated to study their efficiency as a SC. Their excellent surface chemistry via hydrogen bonding improved their structure and characteristics. All the produced aerogels had self-supported 3D nanoporous network topologies with substantial sur-face areas and hierarchical pores. Electrical conductivity tests and the compressive test revealed that these aerogels had an excellent modulus as well as high conductivi-ties. The aerogel was also used as an electrode for SCs, with improved capacitance. Over 74% of the original capacitance was preserved after 1,000 cycles of the cyclic voltammetry test. The aerogel's combination of excellent electrical conductivity, stable nanoporous structure, and BET surface areas may explain the increased elec-trochemical activity. As a result, this aerogel has exhibited a lot of potential for future SC electrode material (Sun *et al.*, 2019).

Compressible and resilient carbon aerogels were fabricated using a low-temperature lyophilization of bamboo cellulose nanofiber solution in liquid nitrogen followed by pyrolysis. Pyrrole was coated onto the carbon aerogels via an *in-situ* polymerization process which resulted in a core-sheath hybrid composite. They showed a high specific capacitance and energy density which after 10,000 cycles retains 88% of its original capacitance. The electrically conductive 3D carbon frameworks and interconnective porous channels of PPy@CA enable efficient fast electron transport, electrolyte ion diffusion, and structural stability. These nanocomposites are promising because of the synergetic impact between high-power-density carbon aerogels and the high energy density of pyrrole (Zhang *et al.*, 2019). Cellulose and graphene oxide (GO) aerogels were fabricated utilizing lithium bromide aqueous solution as the solvent. In situ polymerization of aniline was carried out on these scaffolds to improve their conductivity. They cumulatively exhibited a well-defined 3D porous structure along with good conductivity and high specific capacitance (Li *et al.*, 2020). These materials have the potential to be used in SC applications in the future. Many typical foams, such as PU and polystyrene, pollute the environment, therefore efforts for SC applications have not been concentrated on these materials.

11.5 EMI SHIELDING APPLICATION OF FOAMS

Advances in science and technology have positive as well as negative impacts where electronic devices have contributed high comfort to humans while at the same time causing electromagnetic pollution. These electromagnetic radiations or waves produced cause instruments to malfunction as well as pose a risk to humans. As a result, the eradication of EMI waves has been emphasized as a priority. Two basic approaches can be employed to address this issue, improving distance protection or the use of EMI absorbing materials. The former being impractical the EMI shielding materials have piqued industrial as well as academic interests. Traditional EMI shielding materials include conductive metals and their compounds but their use is limited due to their inherent properties such as low flexibility, high density, narrow absorption bandwidth, corrosion, and high conductivity.

Porous materials have surged in recent research as an excellent excipient of EMI shielding materials due to their lightweight, low thermal conductivity, and adjustable density. A study where acrylic copolymer and silver nanowires foams were fabricated with segregated networks using a high-speed mechanical mixing and hot-pressing method. A drastic increase in electrical conductivity was observed from 0.89 to 142.99 Sm^{-1} with a nanowire volume increase of 0.12 –0.31 vol%. This porous structured high conductive foam surrounded by silver nanowires exhibited an EMI shielding effectiveness (SE) of 63 dB and a specific SE of 180 dB cm^3 g^{-1} in the X-band (8.2–12.4 GHz). The foam broadcasted a low and tunable thermal conductivity by altering the foam density (Liu *et al.*, 2021).

PUFs were studied as a matrix for EMI shielding applications. Their porous structure and surface area enhance their use in these applications. PU has been employed in a variety of technological applications due to its chemical resistance, tensile strength, and processability. Molecular weight, polydispersity, hard and soft segments, and crosslinking ability are commonly used to define the structure of PU. In polyurethane, hydrogen bonding provides a physical relationship between polymer chains, improving overall performance and characteristics. Filler particles in PU foams were used to prepare the core and face skins of sandwich constructions. Stiffness and absorption properties are influenced by the filler particle distribution.

Furthermore, the final material properties are influenced by the filler type, quantity, and thickness. An increase in the porous network along with filler materials improves the reflection of EM waves (Kausar, 2018). Commercial PU foams with a thickness of 10 mm were submerged in a GO and nickel sulfate suspension and compressed repeatedly, followed by a drying process in an air-circulating oven at 90 °C. The addition of Ni enhanced the composite's conductivity and EMI SE substantially. The foams exhibited shielding in the range of 24.03–27.71 dB, which is more than enough to suit the needs of practical applications (Liu *et al.*, 2021). A supercritical CO_2 foaming process was used to create lightweight and flexible thermoplastic PU/reduced GO (TPU/RGO) composite foams for EMI shielding (EMI SE). The hydrogen–bond interaction established in samples due to *in-situ* reduction by L-ascorbic acid contributed to improved surface adhesion and foamed sample flexibility. After $scCO_2$ foaming, multistage cellular was generated, reducing the density of the samples, and allowing the size of the cells to be determined by the RGO concentration. In addition, the construction of segregated structures in TPU/RGO composites considerably improved sample electrical conductivity. Due to the multistage cellular structure and strong conductive network, a shielding efficacy of 21.8 dB was attained with only 3.17 vol% RGO loading. Furthermore, the inclusion of a cellular structure improved the sample's electromagnetic shielding absorption characteristics (Jiang *et al.*, 2019). In another study, microcellular lightweight PU (TPU), Nickel coated carbon nanotubes nanocomposite were prepared by a non-solvent induced phase separation method. These materials showed a higher SE of 69.9 dB (Sang *et al.*, 2021). The EMI shielding efficacy of PU foam materials needs to be investigated further, with an emphasis on filler dispersion and orientation in conductive composites.

Other polymeric foams were also used for EMI shielding applications where PVDF was used as a matrix due to their unique properties such as high-temperature sustainability, chemical resistance, piezoelectric and pyroelectric properties. PVDF and functionalized graphene (FG) composite were evaluated for electrical conductivity and EMI shielding efficacy. The highly conductive nature of FG along with the high aspect ratio foams a conductive network within the polymer matrix which increases the conductivity of the PVDF composite. The conductivity was reported to be directly proportional to the concentration of FG, which indicates that FG is an excellent excipient for EMI shielding applications. By combining polymethylmethacrylate (PMMA) with graphene sheets and foaming with subcritical CO_2 as an ecologically friendly foaming agent, functional PMMA/graphene nanocomposite microcellular foams were created. The incorporation of graphene sheets into insulating PMMA foams improves their electrical conductivity as well as EMI shielding efficiency via an absorption dominating mechanism. Due to the presence of microcellular structures, their ductility and tensile toughness were increased significantly. This work is a promising methodology to fabricate lightweight and tough PMMA-graphene foams with enhanced electrical properties. This is achieved by the combination of functionality and reinforcement characteristics of graphene sheets along with the microcellular structure (Zhang *et al.*, 2011). Polymer foams are excellent recipients for EMI shielding applications due to their lightweight and surface area. Furthermore, their insulating properties inherit their use, which can be modified by the incorporation of conducting fillers.

11.6 SOUND APPLICATION OF FOAMS

Polymer foam's better energy absorption and crashworthiness features have piqued the interest of a variety of industries in recent years, including the automobile and aerospace industries. As unwanted noise has become pollution as well as a health issue, the desire for home safety and a healthier atmosphere has become a fundamental social imperative. Several materials were experimental to address as a material for sound absorption. These materials can be utilized in engineering applications such as room acoustic, industrial noise management, and studio acoustic. PUFs were made using wood fibers as fillers via a crosslinking mechanism of polyol and isocyanate and wood filler. The impedance test, Mettler Toledo density kit, and optical microscope test were done to measure these acoustic properties. The findings demonstrated that 20% polyol with a thickness of 30 mm had the maximum sound absorption coefficient at low and high frequencies, with 0.970 and 0.999, respectively. These results are the outcome of small pore structures and the high density of the fabricated material. When the density is increased the absorption of sound energies also increases which is because the number of fibers in the unit area rises causing surface friction for sound energy (Azahari *et al.*, 2017). The size of the pores in the polymer form composite reduces when the filler % loading is increased. The pores also play a crucial role in sound absorption as these waves travel through these materials dissipating and dampening sound energies. Small pores have more collisions between sound and cell walls which result in a loner path of reflection and refraction, thus a higher sound absorption coefficient.

One of the most fascinating areas in porous material research now is the lowering of pore size in porous materials from the micrometer to the nanoscale scale utilizing up-saleable methods. Two-phase nanoporous foams with a continuous solid phase and a discontinuous or continuous gaseous phase are promising for these applications due to their physical characteristics. Closed-cell microporous and nanoporous PMMA foams were studied for sound absorption application and revealed that nanoporous PMMA has a better absorption coefficient and transmission loss than microporous PPMA (Notario *et al.*, 2016).

Despite their acoustic performance, thermoplastic materials lack appropriate strength, outside weather capabilities, and heat stability. When thermoplastic foams are lit, they shrink, melt, turn into molten droplets, and then help flames. The thermosetting foams, on the other hand, are infusible and insoluble, forming a carbide layer on the burning surface and displaying varying degrees of flame resistance. Verdejo et al. studied flexible polyethylene incorporated with CNTs for acoustic application. He fabricated 5 samples which include blank, 0.01%, 0.05%, 0.1%, and 0.2% (Figure 11.2). An early investigation has demonstrated that even extremely modest CNT loading fractions may significantly improve sound absorption capabilities, as well as the strength-to-weight ratio which is one of the most important properties of such systems (Verdejo *et al.*, 2009). Expandable microspheres can be used to make epoxy foams with adjustable acoustic absorption properties. Modifying the preparation settings impacts the cellular structure from closed to partial open cells, resulting in a variety of sound absorption properties. Short procurement times, high microsphere concentrations, and high foaming temperatures have all been shown to increase

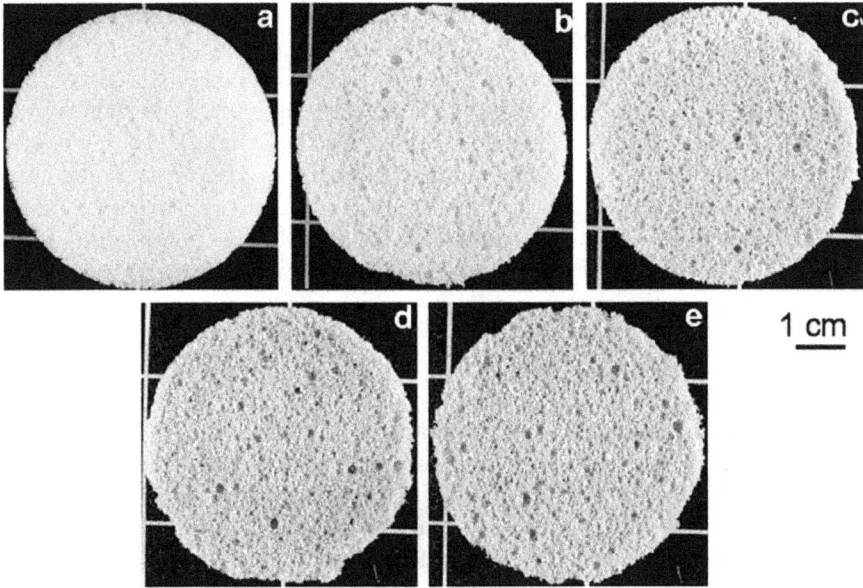

FIGURE 11.2 PUFs (a) Blank, (b) 0.01%, (c) 0.05%, (d) 0.1%, and (e) 0.2%. Reprinted from the reference (Verdejo *et al.*, 2009).

the development of more open cells and improve sound absorption. Finally, a high absorption coefficient of greater than 0.75 will be attained. Novel lightweight epoxy foams can serve as a potential structural-acoustic material, meeting the demands of a wide range of multifunctional systems such as cars, buildings, and aircraft, among others. Even yet, this article just scratches the surface of the subject t, and further research is needed to uncover the underlying processes (Xu *et al.*, 2017).

Aerogels are highly efficient materials for various applications due to their porosity, good pore accessibility, and active sites. For the first time, cellulose/silica aerogels were effectively generated using recycled waste cellulose using precursor methoxy trimethyl silane. Nanocellulose finds enormous applications in the field of energy sectors (Nizam *et al.*, 2021). Their thermal conductivity was 0.04 W/mK on average. Furthermore, the cellulose part of the cellulose/silica aerogels had a 25 °C lower thermal breakdown temperature than cellulose aerogels. They exhibited a sound absorption coefficient of 0.39–0.50 at a thickness of 10 mm which is superior to cellulose aerogels and other commercial polystyrene foams. Silica aerogels are often brittle, but their thermal conductivities are low; cellulose aerogels, on the other hand, may be compressed to an 80% strain, but their thermal conductivities are greater than silica aerogels (Feng *et al.*, 2016). Both cellulose and silica aerogels, on the other hand, are excellent acoustic insulation materials. Furthermore, chitosan aerogels were examined for their acoustic property. The homogenous nanofibrous structure of cross-linked chitosan aerogel generated by supercritical carbon dioxide drying features, mesopores whose main size reduces from 45 to 20 nm as apparent density increases. Regardless of density, the aerogel exhibits peak-shaped sound absorption patterns. The absorption is due to sampling shaking, which reflects the mesoporous structure's exceptionally high flow resistance (Takeshita, Akasaka and Yoda, 2019). An environmentally friendly freeze-drying process was employed to fabricate polyvinyl alcohol, thermally reduced GO (TRGO), and nano clay-based aerogel. The TRGO addition effect on their thermal, physical, and acoustic characteristics was studied. Nanoscale inclusion decreases the pore diameter, whereas the TRGO improved the thermal stability and sound absorption of aerogels (Simón-Herrero et al., 2019). These methods and materials have triggered the research in sound absorption applications.

11.7   EEG APPLICATIONS

EEG is a technique for detecting electrical activity in the brain by placing electrodes on the scalp. It is a strong non-invasive instrument with high temporal resolutions that directly reflects the brain's dynamic processes. It is frequently employed in medical diagnostics and neurobiological studies. Wet electrodes are the most traditional materials often utilized for EEG readings. To lower the skin-electrode contact resistance, wet electrodes require the preparation of skin as well as conduction gel. Conduction gels would invariably leave residues on the scalp, making these treatments inconvenient for users. A short circuit can happen with too much gel or wet electrode being pressed hard onto the scalp which seeps out EEG electrodes. Skin treatment typically entails the outer skin layer abrasion which is time-consuming and painful for patients. Moreover, this has the chance of allergic reactions and infections in patients. The EEG signal quality deteriorates with time as the skin regenerates and as the gel dries. There are however other issues, such as hairs covering the desired measurement spot, resulting in the inadequate skin-electrode contact area. Several types of dry electrodes have been created to overcome the limitations of traditional wet electrodes.

To address the difficulties, a unique dry electrode foam was constructed using an electrically conductive polymer foam that was wrapped by conductive fabric. They displayed both conductivity and polarization which can be used to monitor biopotential. Furthermore, the foam allows for great geometric conformity between the scalp surface and electrode which results in a small electrode-skin impedance even while the electron is in motion. The dry electrode functions better than the wet foam electrode for long-time EEG measurement and is more practical for everyday use (Lin et al., 2011).

Disposal Ag/AgCl electrodes are the common traditional electrode used in most of the conventional EEG measurements. The dielectric layer of the outer layer of the skin called the stratum corneum reduces the transfer mechanism from ions to electrons. For this reason, Ag/AgCl electrode is used as a wet electrode with the aid of a conduction gel, which makes it ion conductive by hydrating the skin's outer layer. To address textile-based electrodes, conducting PUFs were fabricated by incorporating conducting polyaniline onto PU. The surface resistance and impedance of the conductive PU foam were 7 KΩ/square and 1.45 MΩ. Using practical hospital equipment, studies have shown that these PU foam electrodes depict results as compared to that of Ag/AgCl electrodes. For continuous monitoring, these electrodes are promising substituents to wet electrodes (Muthukumar, Thilagavathi and Kannaian, 2016).

The electrode for hairy places is the most difficult aspect of wearing an electroencephalogram. Wet electrodes are yet another alternative, but they require conductive gels, that dry out with time and are therefore unsuitable for long-term monitoring. To address these issues a multilayer and flexible electrode for EEG monitoring was proposed. A hybrid material comprised of PDMS as matrix and silver nanoparticles with a 40% fraction as filler was fabricated. The silicone elastic substance PDMS is biocompatible and easy to produce. The probe structure of the electrode enhances its efficiency in hairy areas. Electrolytes are stored in a reservoir layer which is released via a foam layer during the long period of EEG monitoring, to the scalp–electrode interface. This ensures a low scalp-electrode contact impedance (Hua *et al.*, 2019). These multilayer flexible electrodes can be employed for future EEG monitoring applications. 3D-printed PDMS/Ag-based foams were also evaluated for the EEG applications and showed promising results (Wang *et al.*, 2018). Many polymer-based composites are being researched in this area due to the tunability and tailoring properties of polymers.

11.8   POLYMER FOAMS IN ENERGY ABSORBING APPLICATION

The compression behavior of foamed polymers is used in the construction of energy-absorbing structures, which is a significant commercial application. A foamed polymer is appropriate for such applications because it may regulate load-compression response by varying cell shape, density, and matrix. Energy-absorbing materials are often chosen through empirical, trial-and-error processes rather than analytic methodologies as information connecting the energy-absorbing characteristics to the foam factors is not always accessible. These interactions must be quantified to attain the best design and material (Rusch, 1970). Polymer foams have a cellular structure that imparts superior insulating and impact properties. Energy impacts on polymeric foams are spread in such a manner that the maximum force is always below a specific limit. This is due to the deformation mechanism of the structure of the cell. During an impact, the maximum force acting on a polymer foam is substantially below the maximum force that acts upon unidentical non-foamed materials. Therefore, foams are being employed in engineering applications. In the transportation sector, foams are widely employed for passenger safety. Their impact-damping properties are used for packing and other applications, to exemplify the protect the product during transit, etc. Similarly, the safety of athletes and their ability to prevent damage is critical, which is why diverse foam structures are utilized in a variety of applications, which include protective gears and helmets. Other sports activities such as the high jump, pole vault, and gymnastics employ mats to cushion landing (Tomin and Kmetty, 2021).

Crashworthiness testing is an important criterion in aircraft structural design, and it can assist to lower the risk of damage and death in possibly avertable crashes. Different crashworthiness design concepts have been created as a result of the arrangement of energy absorption components and the process of energy absorption being linked to the type of aircraft. The requirement to protect people from extreme loading conditions, such as those associated with quick decelerations or external impact, is becoming increasingly critical as the need for high-speed, energy-efficient transportation systems grows. Many attempts have been made to build crash-worthy buildings, including the use of composite or metallic tube sections at strategic locations throughout the design. It is commonly known that tube-like structures may absorb a significant amount of energy when loaded in an axial direction if designed and used correctly. Polymer foam blocks are crucial for energy absorption, particularly during the early impact period. The area for the cargo hold has been seen to be conserved by putting crushable Rohacell-31 foam blocks into the space between the cargo floor and the belly skin. More than 70% of the maximum energy absorption by the foam block occurred in the first 50 ms, and energy absorption by the foam block becomes more efficient as the foam block length declines (Zheng *et al.*, 2011). Lightweight energy-absorbing materials were cast out of PVC foams filled with metal tubes. The findings reveal that aluminum tubes have a higher energy-absorbing capacity than steel tubes and that the specific energy absorption (SEA) remains relatively constant as the tube length of metal increases. The ability of the two types of metal tubes to absorb energy rose as the inner diameter/thickness ratio decreased (Alia *et al.*, 2015).

A tube's energy capacity is determined by its SEA value, that is total crushing energy absorbed per unit material. A variety of lightweight bamboo-reinforced foam constructions were investigated for their energy-absorbing properties. The initial focus is on defining the energy-absorption properties of bamboo tubes and determining how tube shape affects energy absorption. Due to axial splitting associated with barreling of the samples under compressive stress, the SEA of the bamboo tubes decreases with increasing tube length. The effect of the tube inner diameter/thickness (D/t) ratio was studied using a variety of tube diameters, and it was observed that reducing D/t results in a slight increase in SEA (Umer *et al.*, 2014).

The above-mentioned are some of the areas where polymer foams are being employed for applications. Albeit, polymer foams are a threat to the environment they cannot be completely eradicated due to their extensive application in many sectors.

11.9 CONCLUSION

Polymer foams are commercial products in different applications such as bedding, and shock absorption, as well as in the electronic sector. Their increasing demand cannot be halted but alternate methods to enhance their biodegradability can be addressed. Different attempts are being researched to improve biodegradability by using natural-based monomers for their production. Their importance in the energy sector is discussed in this chapter which talks about SCs, EMI shielding, EEG application, acoustic, and energy absorption applications. The excellent properties such as high porosity and surface area enhance this material in many applications.

REFERENCES

Alia, R. A. *et al.* (2015) 'The energy-absorption characteristics of metal tube-reinforced polymer foams', *Journal of Sandwich Structures and Materials*, 17(1), pp. 74–94. doi: 10.1177/1099636214554597.

Azahari, M. S. M. *et al.* (2017) 'Acoustic properties of polymer foam composites blended with different percentage loadings of natural fiber', *IOP Conference Series: Materials Science and Engineering*, 244(1). doi: 10.1088/1757-899X/244/1/012009.

Cateto, C. A. *et al.* (2014) 'Lignin-based rigid polyurethane foams with improved biodegradation', *Journal of Cellular Plastics*, 50(1), pp. 81–95. doi: 10.1177/0021955X13504774.

Feng, J. *et al.* (2016) 'Silica–cellulose hybrid aerogels for thermal and acoustic insulation applications', *Colloids and Surfaces A: Physicochemical and Engineering Aspects*, 506, pp. 298–305. doi: 10.1016/j.colsurfa.2016.06.052.

Ginty, P. J. *et al.* (2008) 'Controlling protein release from scaffolds using polymer blends and composites', *European Journal of Pharmaceutics and Biopharmaceutics*, 68(1), pp. 82–89. doi: 10.1016/j.ejpb.2007.05.023.

Gong, P. *et al.* (2018) 'Environmentally friendly polylactic acid-based thermal insulation foams blown with supercritical CO_2', *Industrial and Engineering Chemistry Research*, 57(15), pp. 5464–5471. doi: 10.1021/acs.iecr.7b05023.

Gu, R., Konar, S. and Sain, M. (2012) 'Preparation and characterization of sustainable polyurethane foams from soybean oils', *JAOCS, Journal of the American Oil Chemists' Society*, 89(11), pp. 2103–2111. doi: 10.1007/s11746-012-2109-8.

Hua, H. *et al.* (2019) 'Flexible multi-layer semi-dry electrode for scalp EEG measurements at hairy sites', *Micromachines*, 10(8), pp. 1–13. doi: 10.3390/mi10080518.

Jang, S. *et al.* (2021) 'A hierarchically tailored wrinkled three-dimensional foam for enhanced elastic supercapacitor electrodes', *Nano Letters*, 21(16), pp. 7079–7085. doi: 10.1021/acs.nanolett.1c01384.

Jiang, Q. *et al.* (2019) 'Flexible thermoplastic polyurethane/reduced graphene oxide composite foams for electromagnetic interference shielding with high absorption characteristic', *Composites Part A: Applied Science and Manufacturing*, 123(May), pp. 310–319. doi: 10.1016/j.compositesa.2019.05.017.

Jung, P. U. (2013) *Development of Innovative Gas-Assisted Foam Injection Molding Technology*, pp. 1–217.

Kausar, A. (2018) 'Polyurethane composite foams in high-performance applications: A review', *Polymer - Plastics Technology and Engineering*, 57(4), pp. 346–369. doi: 10.1080/03602559.2017.1329433.

Li, Y. *et al.* (2020) 'Green synthesis of free standing cellulose/graphene oxide/polyaniline aerogel electrode for high-performance flexible all-solid-state supercapacitors', *Nanomaterials*, 10(8), pp. 1–18. doi: 10.3390/nano10081546.

Lin, C. T. *et al.* (2011) 'Novel dry polymer foam electrodes for long-term EEG measurement', *IEEE Transactions on Biomedical Engineering*, 58(5), pp. 1200–1207. doi: 10.1109/TBME.2010.2102353.

Liu, F. *et al.* (2021) 'Electromagnetic interference shielding property of silver nanowires/polymer foams with low thermal conductivity', *Journal of Materials Science: Materials in Electronics*, 32(24), pp. 28394–28405. doi: 10.1007/s10854-021-07219-0.

Miao, S. *et al.* (2012) 'Soybean oil-based polyurethane networks as candidate biomaterials: Synthesis and biocompatibility', *European Journal of Lipid Science and Technology*, 114(10), pp. 1165–1174. doi: 10.1002/ejlt.201200050.

Muthukumar, N., Thilagavathi, G. and Kannaian, T. (2016) 'Polyaniline-coated foam electrodes for electroencephalography (EEG) measurement', *Journal of the Textile Institute*, 107(3), pp. 283–290. doi: 10.1080/00405000.2015.1028248.

Nizam, P. A. *et al.* (2021) *Nanocellulose Based Composites for Electronics*. Elsevier Inc. doi: 10.1016/b978-0-12-822350-5.00002-3.

Nofar, M. and Park, C. B. (2014) 'Poly (lactic acid) foaming', *Progress in Polymer Science*, 39(10), pp. 1721–1741. doi: 10.1016/j.progpolymsci.2014.04.001.

Notario, B. *et al.* (2016) 'Nanoporous PMMA: A novel system with different acoustic properties', *Materials Letters*, 168, pp. 76–79. doi: 10.1016/j.matlet.2016.01.037.

Rusch, K. C. (1970) 'Energy-absorbing characteristics of foamed polymers', *Journal of Applied Polymer Science*, 14(6), pp. 1433–1447. doi: 10.1002/app.1970.070140603.

Sang, G. *et al.* (2021) 'Interface engineered microcellular magnetic conductive polyurethane nanocomposite foams for electromagnetic interference shielding', *Nano-Micro Letters*, 13(1), pp. 1–16. doi: 10.1007/s40820-021-00677-5.

Sauceau, M. *et al.* (2011) 'New challenges in polymer foaming: A review of extrusion processes assisted by supercritical carbon dioxide', *Progress in Polymer Science (Oxford)*, 36(6), pp. 749–766. doi: 10.1016/j.progpolymsci.2010.12.004.

Shah Mohammadi, M., Bureau, M. N. and Nazhat, S. N. (2014) *Polylactic acid (PLA) biomedical foams for tissue engineering, Biomedical Foams for Tissue Engineering Applications*. Woodhead Publishing Limited. doi: 10.1533/9780857097033.2.313.

Simón-Herrero, C. *et al.* (2019) 'PVA/nanoclay/graphene oxide aerogels with enhanced sound absorption properties', *Applied Acoustics*, 156, pp. 40–45. doi: 10.1016/j.apacoust.2019.06.023.

Sun, Y. *et al.* (2019) 'Synthesis of polypyrrole coated melamine foam by in-situ interfacial polymerization method for highly compressible and flexible supercapacitor', *Journal of Colloid and Interface Science*, 557, pp. 617–627. doi: 10.1016/j.jcis.2019.09.065.

Takeshita, S., Akasaka, S. and Yoda, S. (2019) 'Structural and acoustic properties of transparent chitosan aerogel', *Materials Letters*, 254, pp. 258–261. doi: 10.1016/j.matlet.2019.07.064.

Tomin, M. and Kmetty, Á. (2021) 'Polymer foams as advanced energy absorbing materials for sports applications—A review', *Journal of Applied Polymer Science*, (August 2021), pp. 1–23. doi: 10.1002/app.51714.

Umer, R. *et al.* (2014) 'The energy-absorbing characteristics of polymer foams reinforced with bamboo tubes', *Journal of Sandwich Structures and Materials*, 16(1), pp. 108–122. doi: 10.1177/1099636213509644.

Verdejo, R. *et al.* (2009) 'Enhanced acoustic damping in flexible polyurethane foams filled with carbon nanotubes', *Composites Science and Technology*, 69(10), pp. 1564–1569. doi: 10.1016/j.compscitech.2008.07.003.

Wang, Z. *et al.* (2018) 'A multichannel EEG acquisition system with novel Ag NWs/PDMS flexible dry electrodes', *Proceedings of the Annual International Conference of the IEEE Engineering in Medicine and Biology Society, EMBS*, 2018-July(2017), pp. 1299–1302. doi: 10.1109/EMBC.2018.8512563.

Xu, Y. *et al.* (2017) 'Epoxy foams with tunable acoustic absorption behavior', *Materials Letters*, 194, pp. 234–237. doi: 10.1016/j.matlet.2017.02.054.

Zhang, H. Bin *et al.* (2011) 'Tough graphene-polymer microcellular foams for electromagnetic interference shielding', *ACS Applied Materials and Interfaces*, 3(3), pp. 918–924. doi: 10.1021/am200021v.

Zhang, L. *et al.* (2013) 'The study of mechanical behavior and flame retardancy of castor oil phosphate-based rigid polyurethane foam composites containing expanded graphite and triethyl phosphate', *Polymer Degradation and Stability*, 98(12), pp. 2784–2794. doi: 10.1016/j.polymdegradstab.2013.10.015.

Zhang, X. *et al.* (2019) 'In-situ growth of polypyrrole onto bamboo cellulose-derived compressible carbon aerogels for high performance supercapacitors', *Electrochimica Acta*, 301, pp. 55–62. doi: 10.1016/j.electacta.2019.01.166.

Zheng, J. *et al.* (2011) 'Crashworthiness design of transport aircraft subfloor using polymer foams', *International Journal of Crashworthiness*, 16(4), pp. 375–383. doi: 10.1080/13588265.2011.593979.

Index

Note: **Bold** page numbers refer to tables and *italic* page numbers refer to figures.

For Product Safety Concerns and Information please contact our EU
representative GPSR@taylorandfrancis.com
Taylor & Francis Verlag GmbH, Kaufingerstraße 24, 80331 München, Germany

www.ingramcontent.com/pod-product-compliance
Lightning Source LLC
Chambersburg PA
CBHW060554220326
41598CB00024B/3106

9 781032 111704